Sensitivity Analysis and Validation of a Novel Fuselage Design

ARVIND

Contents

Contents

Chapter 1

Introduction

1.1 Wide-body Transport Aircraft: Market demand

Air transport is anticipated to grow at a comparable rate in the coming years. Air traffic forecasts an annual rise of 3.7 percent in the coming 15th year, exceeding over 145,000,000,000 revenue passenger kilometers (RPK), far beyond doubling from 2016 [86]. Statistically, the increase in air traffic or fleets leads to connects its adverse footprint from noise, fuel burn, and, more significantly, emissions [88] [92] [6]. The "IATA" initiated and addressed a significant approach towards sustainable targets with a "Four-pillar Strategy" [86] to meet the carbon footprint targets represented in Figure 1.1. The four key strategy factors are intertwined in a manner that enables the achievement of sustainable goals. Such a strategic pillar not only aims to reduce emissions but also provides better aircraft performance. Meanwhile, all airline industrialists are bound to agree on optimistic carbon reduction objectives involving improving fuel economy by 1.5 percent per year between 2009 and 2020 and reducing net annual aviation carbon footprint by half by 2050 compared to 2005 [86]. According to fleet analysis, accounting for aircraft configuration is vital when using metrics like CO2 per seat per mile [88]. The influence

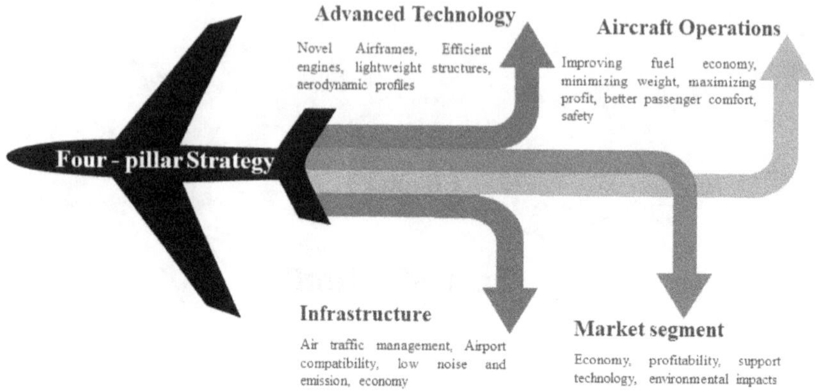

Figure 1.1: Four pillar strategy to reduce aviation carbon footprint

of emerging concepts of aircraft design can potentially cause environmental impacts as estimated over distinct parameters representing differing levels of configuration, air traffic volume, and structural integrity [ ] [ ]. Differences in seating class and density capacity result in varying seat counts per aircraft, impacting the CO2 emissions per seat mile, with higher-density planes yielding a more unfavorable emission. In the coming years, the growth rate in terms of the number of widebody jets is going to hike by more than 51 percent as shown in Figure 1.2. IATA depicts an investigation of aircraft type and the intensity of carbon footprints from 2013 to 2019 in Figure 1.3. It outlines that the majority of the emission and environmental impacts are caused by narrow-body aircraft. Wide-body is capable of fleeting long haul range and efficient in SFC and leads less emission to the environment. ICAO enforced rules and certifications to improve operational and infrastructure efficiency. Future aircraft can ignite more initiation to neutralize the carbon offset.

In 2015, Boeing harnessed the influential factors of high-density capacity and introduced a market niche for a new airplane configuration called middle-of-the-market (MoM) air-

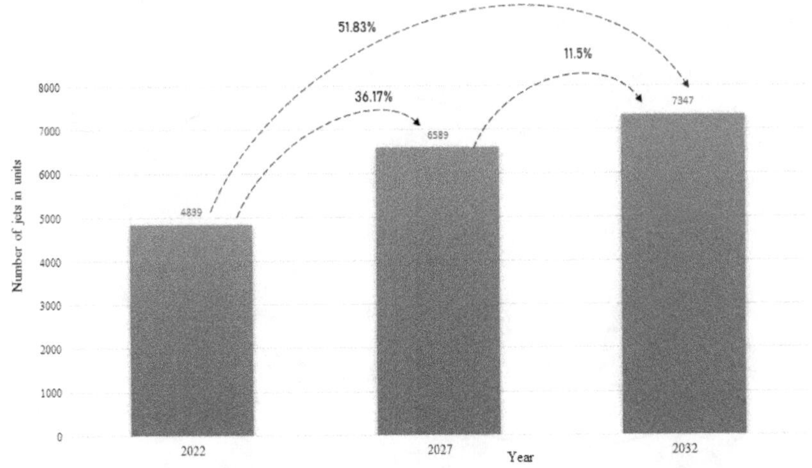

Figure 1.2: Growth of number of widebody jets in coming years

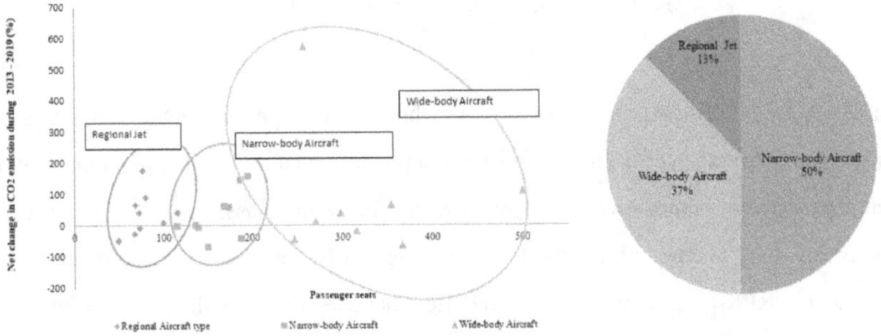

Figure 1.3: Percent of CO2 emission rate by aircraft type from 2013 to 2019

craft. A Middle of the Market (MoM) transport aircraft serves as a bridge between current and upcoming single-aisle and smaller dual-aisle planes, offering optimized capacity and range [91] [49]. The concept of MoM was gleaned from an advanced twin-aisle medium range (ATMR) from 5,000 km to 9,500 km and a seating capacity of 180 to 300 [53] [49] are illustrated in Figure 1.4.

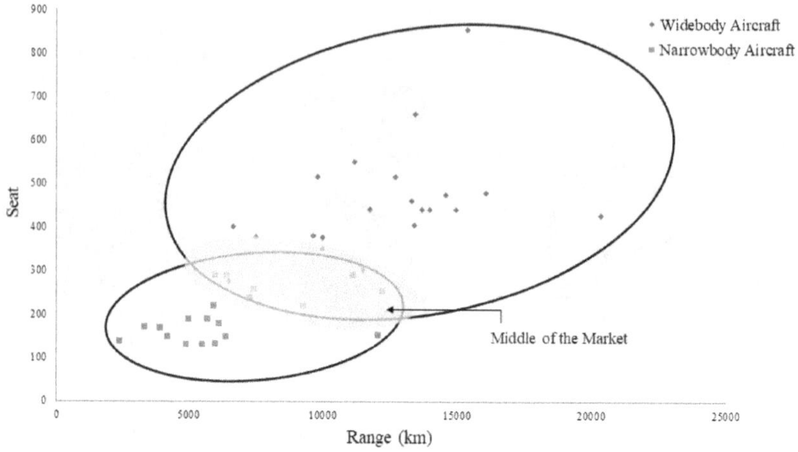

Figure 1.4: The Middle-of-the-Market aircraft zone

In 2019, the ICCT assessed emissions and carbon intensity per passenger, accounting for seating and aircraft types. The report shows depending on aircraft type, premium seating emitted 2.6 to 4.3 times more CO2 per RPK than economy class. It states that the wide-body economy class exhibited the lowest average carbon intensity at approximately 65 g CO2 per RPK [illegible]. Johnsson [illegible] reported that executing a wide-body configuration can produce 30 percent more revenue and 40 percent lower transit expenses than narrow bodies [illegible]. Airbus CEO Guillaume Faury anticipates a surge in demand for wide-body aircraft shortly, particularly in 2023 and 2024, as stated during an event organized by AJ-PAE on December 8, 2022, as reported by Reuters [illegible]. The current wide-body aircraft market reflects the future potential, with GE Aerospace CEO stating India could add 130-150 wide-body planes for domestic routes [illegible]. There is evident potential demand for wide-body aircraft, as indicated by recent aviation news. Several Airlines currently deploy Boeing 737, Airbus A320, Boeing 757, Boeing 767, Airbus A310, and Airbus A330 aircraft on MoM routes. But the pandemic indeed influenced the global gross domestic product (GDP) and posed a threat to the airline's revenue. The latest study says the global aircraft manufacturing market (GAMM) will be worth USD 626.25 billion by 2030. The report forecasts for future market growth demand are reviving by increasing production to meet resurgent demand for wide-body jets [illegible] .

1.2 Emerging Programs and Innovations in wide-body aircraft

1.2.1 Programs

The International Civil Aviation Organization (ICAO) postulated strict sustainability targets for the new transport aircraft, significant reduction in emissions and enhancement in fuel economy. Air transport design technology is not just about enriching fuel efficiency, but also a vital aspect of boosting flight ground services, providing passenger comfort, and minimizing environmental consequences. Many R and D, funded programs have taken the initiative to reduce the carbon footprints that help to increase passenger mobility, comfort, economic measures, and structural integrity. Well-popular programs began neutralizing the carbon footprints such as CORSIA, ACARE, ERA, CLEEN [illegible] [illegible], GARDN, Clean Sky 2 [illegible], ICCT [illegible] and ICAO Programme. Such programs offer a reduced CO2 plan such as the idea of operating in increasingly economical planes, making better use of emerging technology, using green technologies to reduce emissions, and decreasing air traffic congestion. Such logistical programs to mitigate climate change provide a win-win scenario [105] [169] [86][66].

1.2.2 Current Trends

The most recent airplane design technologies and programs mark a substantial advancement in aviation. They not only offer more efficient and environmental aviation travel, but they also open up fascinating future possibilities. The introduction of innovative techniques and progressive designs is at the heart of advanced airframe design. The first pillar is design technology which comprises short-term, medium-term, and longer-term including retrofit and upgraded production, new airframe designs, and radical airframe technologies. Aerodynamics, propulsive framework, and retrofit systems are the most significant ways to improve design effectiveness in the aviation industry briefly described below.

Aerodynamics

The aerodynamic concept has rapidly grown resulting in innovations featuring considerably lower drag, SFC, and emission. Since the 1970s, winglet design in aircraft has shown a significant result by reducing the drag reduction technique [78]. Raked wingtip and spiroid wingtip can be significantly capable of reducing drag and fuel burn up to 6 percent [123] [146]. Simultaneously a unique feature a variable camber wing came into aerospace research. Variable camber wing distinctly improves aerodynamic efficiency (L/D) and reduces drag, SFC, and noise by 2 percent [illegible]. Currently, morphing systems like ATE devices are applied in the wing to lower the SFC by 2.5 percent [179] [65]. Research further extended to riblets to trigger shear stress reduction by 16 percent. Riblets are potential in all types of flow regimes and can reduce fuel burn by 1 percent [illegible] [illegible].

Boeing collaborated with the CLEEN II program to save fuel burn. R and D developed SEW for new aircraft design and to reduce weight without compromising fuel efficiency by 4 percent [66] [86] [170]. Eminent researcher and scientist from NASA investigating SEW and Spanwise adaptive wings using shape memory alloy [100]. This mechanism improves the aerodynamic performance enabling a reduction of the SFC by 8 percent [65]. LFC is another aerodynamics technique explored for several generations and lately achieved advances throughout the design. LFC is presently one such technique that can significantly increase fuel efficiency by greater than 30 percent [86] [40]. Few case studies are studied on double bubble aircraft [197] [198], joint wings [86], blended wing body (BWB) [100], Flying V aircraft [178] [181], and strut-braced wings [38] [36]. A common reason for fuel-efficiency benefits due to drag minimization, better aerodynamic profile, improved lift, and weight reduction are illustrated in Table 1.1.

Table 1.1: Aerodynamics Contribution

Aerodynamic Configuration	**Benefits**	**Sources**
Winglets	Reduce tip vortices; Induced drag reduction	[186] [78] [179]
Vortex generators	Delay flow separation; Improve lift	[186] [65]
Riblets	drag-reduction	[35] [122] [186]
Variable camber	Reduce wave drag	[40]
LFC	Robust laminar flow; Delay stalling; Lower drag;	[86] [40]
HLFC	Drag reduction; Decreases power consumption; Improve fuel-efficiency	[162]
AFSC	Drag-reduction; Noise reduction;	[162]

Propulsive systems

The new engine technology and notable engine components integrate fuel burn reduction. Currently, turbofan engines operating more than nine bypass ratios (BPR) initiate to reduce SFC. RR trend 7000 and GE9X engines provide 10 BPR to produce 10 percent fuel efficiency [195].

Table 1.2: Contribution on Propulsive systems

Propulsive Systems	Benefits	Sources
New engine architecture	Reduce fuel burn; noise reduction	[195]
UHBR engine	High BPR; Lower fuel burn; Improved SFC; Reduction in noise.	[74] [130]
BLI	Drag-reduction; Structural integrity; Better fuel consumption.	[34] [84]
Advance and Ultra Fan Engine	Low emission. Fuel Efficient	[74] [130] [84]
Open rotor Engine	Fuel efficient	[139] [145]
Noise suppression technology	Redesign in the airframe drag-noise reduction	[121] [86]
Low-emission technology	Increase in BPR; Sizing fan size; Improved SFC; improves combustion efficiency; better structural integrity.	[illegible] [86] [illegible]

Rolls-Royce and Safran have experimented with Advance and Ultrafan engine and Ultra-High-Bypass Ratio (UHBR) turbofan engine forecasts to achieve 20 to 25 percent less SFC over RR Trend 800 [74] [130]. Another new variant of turboprop engines is a propfan or open rotor. New engine architecture like open rotor and GTF engine is a massive future demand. The open rotor is a combination of propeller and turbofan engine that enables a 30 percent lower SFC compared to CFM56 [139]. A SUGAR program initiated with NASA and Boeing designed a novel aircraft with an open rotor. Open rotor por-

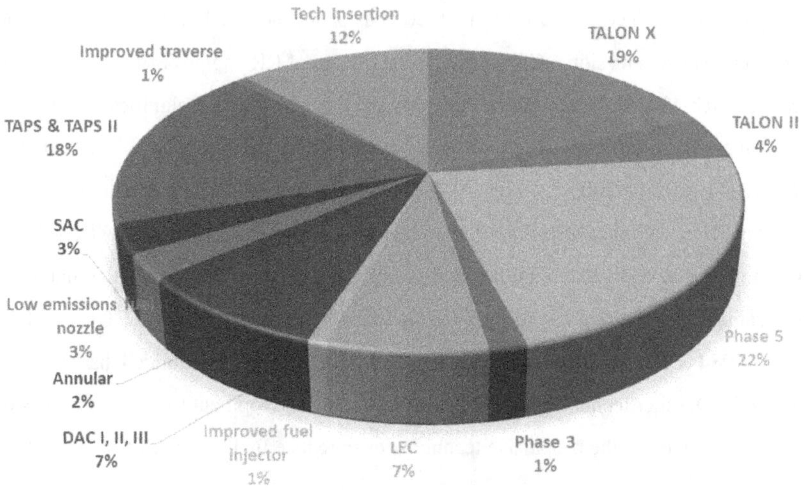

Figure 1.5: Different combustor technologies

traits challenge severe noise levels [ ] and limited speed range up to 0.8 Mach speed. Whereas the GTF engine is combined with a gearbox with a fan which produces high BPR leading to reduced fuel burn significantly [ ]. In June 2021, GE and Safran collaborated and initiated a CFM RISE Program to reduce carbon footprint by 20 percent [ ]. Collins Aerospace's innovative clean fan duct and short inlet can significantly improve 1 percent fuel economy and help in noise reduction by 2.0 EPNdB [ ].In addition, BLI technology is a possible way to reduce power requirements, lower carbon emissions, and improve fuel economy [ ] [ ]. As per studies, the majority of jet engines emit 70 percent CO2 exhaust gases. Most airplane noise is from airframe structure, fan, flap, and jet [ ]. Many researchers and engineers are working on jet noise suppression by altering airframe integration (Shape optimization, chevon, duct flow, Micro-tab device, Acoustic liners) or adaptive Flow Control (Active noise control, Acoustic boundary layer). Noise suppression devices can reduce noise propagation and operating costs [ ]. Rolls-Royce

Corporation has incorporated different Combustor technology like type 1, type 2, and type 3 to reduce emissions. Currently, the R&D team of RR engines is introducing Rich-Burn, Quick-Mix, and Lean-Burn (RQL) combustors which have a significant method for reducing NOx emissions. There are some trending RR engines such as Trent7000-68, Trent7000-70, Trent7000-72, Trent7000-72C, and Trent7000-72D, specially designed to improve Non-volatile Particulate Matter (nvPM). Furthermore, the different combustor technology was analyzed and illustrated in Fig 1.5. Figure 1.5 represents that maximum utilization of technology was followed in the following fashion as Phase 5 RQL > TALON X > TAPS I & II > Tech Insertion > DAC I, II, & III, > LEC. It was noticed that due to on-going R&D, other technologies were in use or under development to emit less emission. The overview of all the propulsive technologies to reduce fuel burn is illustrated in Table 1.2.

Advanced Structures and Materials

The aviation industry has gained responsibility for the implementation of composite processes and production technologies. The potential demand to improve the high strength-to-weight ratio tends to decrease SFC at a range of flights. Weight reduction triggers to reduce the SFC and improve performance measures, fuel economy, and competitiveness. The most common material like composite material, lightweight alloys, and advanced materials capable of reducing the gross weight, finally leads to decreasing fuel burn of the aircraft. Most well-known composite components like belly fairings in A380, rear pressure bulkhead in A340-600, floor beams in A350, B787, Keel beam in A340-600, nose cone, fan blades, nacelle, and many more. Further weight reduction is possible from advanced systems. The aircraft interior attempts to balance elegance, security, and passenger leisure with operational performance measures. New ideas are coupled with lightweight performance and strict quality assurance signifying fuel economy by 5 percent [86]. Safran proved EGTS drastically saves fuel consumption on the phase of taxiing on the LTO cycle. As a result, engine architecture and its components like the combustor, engine cooling system, and rotating system are the driving constraints to fuel consumption and carbon emission.

1.2.3 Emerging Concepts

A complete insight into how a combination of technologies in aircraft design can neutralize carbon footprints and reduce fuel burn impacts is shown in Figure 1.6. Future aircraft can neutralize the carbon offset based on the utilization of sustainable approaches summarizes the facts and figures Figure 1.6.

A combination of technologies in aircraft design can neutralize carbon footprints and reduce environmental impacts. Subsonic Ultra Green Aircraft Research (SUGAR) Program initiated future aircraft using the open rotor [28] [129]. Such a program implemented an ultra-thin strut-braced wing concept to eliminate engine emission and acoustic causes [174]. Warwick (2021) reported that Bauhaus Luftfahrt initiated a novel innovation Propulsive Fuselage Concept (PFC). Implementing boundary layer ingestion (BLI) triggers to minimize weight and drag [163]. A similar progressive concept designed by the EU with Bauhaus Luftfahrt is the propulsive fuselage concept (PFC) [77] [50]. PFC provides minimized drag, a tail-mounted fan collects flow, and increases the BPR leading to improved engine efficiency [156] [145]. A similar kind of BLI application in Aurora D8 aircraft causes a reduction of 35 percent SFC approximately [3]. BLI improves the performance, SFC, mass flow ratio, and efficient rotating components of an engine [30] [84]. IATA's [86] report states that electric and hybrid electric flying vehicles will begin to operate from 2035 to 2045. The benefits of electrically powered aircraft release zero-emission, but energy density and performance are still challenging tasks to pull up.

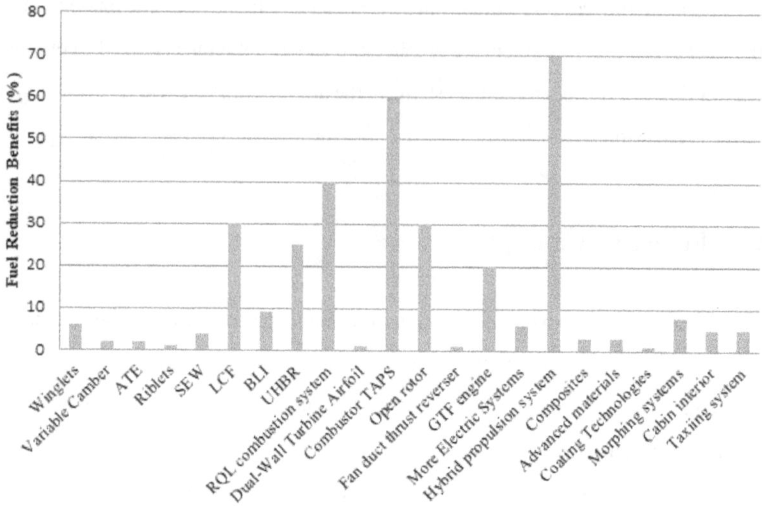

Figure 1.6: Innovative technology enhance fuel efficiency in percent

Zunum Aero, E-Fan X, and STARC-ABL electric airplanes with 50 capacities will enter the airline service in 2027 [ ] [ ].

In a nutshell, the overall design unit is moderately influenced by aerodynamic consideration techniques for flight performance; Slight to moderate impact on cost and environmental issues [ ]. The implementation of airframe design technology as an aerodynamic profile by drag reduction, increasing BPR, lightweight structure, and disruptive techniques trigger to the reduction of fuel consumption and improves performance is tabulated in 1.3 [ ]. Revolutionary aircraft designs such as double bubble aircraft [ ], Flying V aircraft, and Blended wing Body aircraft [ ] design significantly improve fuel efficiency by more than 30 percent in coming years are illustrated in Figure 1.7. Fur-

Figure 1.7: Typical airframe technologies

thermore, a comprehensive understanding of the different program scopes to improve the economic benefits and reduce the environmental issues that arise from aviation noise and emissions is shown in Figure 1.8. Such schematic layout(Figure 1.8) represents different design technologies that are yet to be implemented to trigger sustainable development in terms of percentage.

1.3 Research Motivations

Air transport has grown into continuous progress in technological innovation to improve fuel efficiency and lower carbon footprint. Advanced technology proved to lower fuel

Table 1.3: Airframe technologies

Airframe technologies	Benefits	Sources
Strut-braced wing	High AR; Lowered drag; fuel efficiency; Long-range.	[28] [129]
BWB	Increase the payload capacity; Lifting body; lowers drag; Fuel-efficient	[106]
Double bubble fuselage	Increased passenger comfort; lower structural weight; efficient engine; less noise and emission Suitable for airport compatibility	[illegible] [illegible]
PFC	Reduction of wake generation; reduced drag.	[illegible] [illegible]
Flying V	Improve passenger capacity; long-range; lowers fuel burn; Airport compatibility	[illegible]
Hybrid-Electric Aircraft	Low emission; no pollutants; lower noise; fuel efficiency	[illegible]

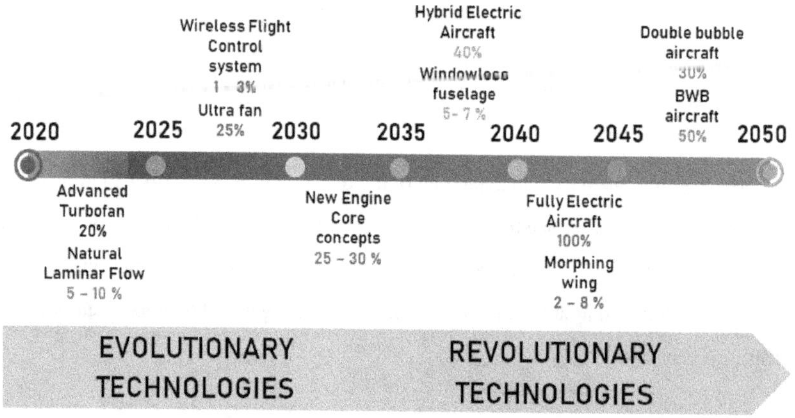

Figure 1.8: Airframe technologies future framework

consumption in many ways by improving flight ground services, providing passenger comfort, and reducing environmental effects [125] [145]. It is expected that wide-body is going to be dominant in the coming aviation sector. In an era where efficiency is not only a goal but a need, optimizing the airplane fuselage becomes critical. This initiative isn't just about making airplanes more efficient and agile; it's about conceptualizing a future in which every ounce of material, every contour, and every structural component works in tandem to produce a performance masterpiece [68]. The pursuit of excellence in aircraft design has always been about pushing the limits of what is possible. "Design Optimisation of Aircraft Fuselage for Enhanced Performance and Efficiency" contains a vision in which creativity and engineering collide to alter the very essence of flying. We uncover the mysteries of the fuselage's shape as we dig into the complexities of aerodynamics, materials science, and cutting-edge engineering. Each improvement has the potential to improve performance, reduce fuel consumption, and pave the door for more environmentally friendly aircraft solutions. "Design Optimisation of Aircraft Fuselage for Enhanced Performance and Efficiency" is more than a chore; it's a transformational goal. It's an appeal to engineers, inventors, and visionaries to work together to take aviation to new heights, inspiring future generations. This isn't just research; it's a manifesto for the future of flying, promising to make the unthinkable attainable.

1.4 Objectives of the study

The proposed objectives of the research are:

i. Analysis of influential factors affecting wide-body fuselage.

ii. Selection of wide-body aircraft and comparative analysis of geometry, aerodynamics performance, and human factor issues.

iii. Decision-making on optimal fuselage design through Multi-Criteria Decision-Making Techniques and Multidisciplinary Design Optimization Techniques.

iv. Validating the Optimum design with airworthiness standards and Wind tunnel Experimentation.

Chapter 2

Statistical Analysis of Transport Aircraft for Identifying Influential Factors

2.1 Background

Airplane design is not a one-time event but a continuous, iterative process [144]. Designers and engineers build upon the knowledge and experience gained from previous aircraft designs. Each new design draws from the lessons learned from past successes and challenges, incorporating improvements and innovations to create more efficient and advanced aircraft. Existing airplanes, especially those that share similarities with the proposed design, serve as valuable sources of data [132]. This data provides crucial guidance for initiating the design process. Designers can study the performance, aerodynamics, materials, and technologies used in existing aircraft to set the foundation for the new design. This knowledge helps in establishing initial design values and parameters [63]. Aircraft statistical data is not only valuable for ensuring that the aircraft meets safety and regulatory standards such as FAA but also for understanding the overall performance. These areas may include fuel efficiency, emissions reduction, structural integrity, safety, and passenger comfort. By analyzing this data, designers can pinpoint specific aspects that need improvement.

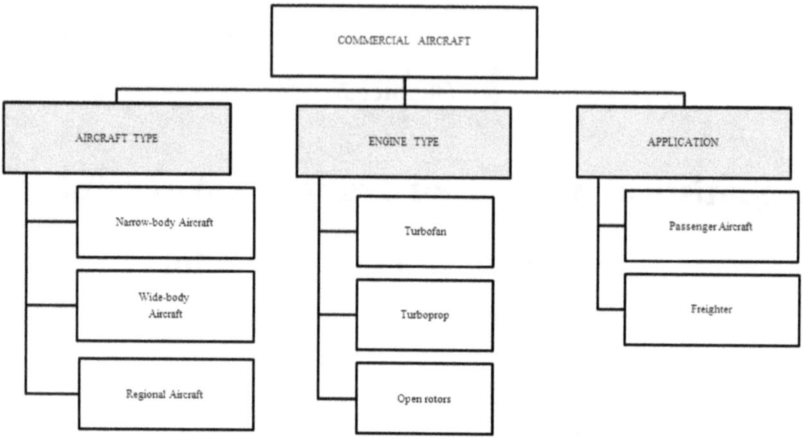

Figure 2.1: Classification of commercial airplane

Today's transport aircraft are available in a wide range of designs. Several new passenger-based aircraft have recently initiated commercial service, and yet more to come in the coming years. According to the 14 CFR part 25, transport category airplanes can be classified into three groups: (a) the aircraft types, (b) engine types, and (c) applications, either passenger or freighter, as illustrated in Figure 2.1. Any aircraft design begins with mission requirements based on the operation and purpose of the operation. Throughout an aircraft's serviceability, its responsibilities and operations will evolve considerably beyond those expected when the initial concept criteria were originally set are represented in Table 2.1 Airframe criteria come in a variety of kinds though from a diversity of ways. There are also different "customer-centric" criteria, which include the real purpose and operational skills required by new buyers [144] [109] [164]. It included apparent criteria such as distance, capacity, and velocity, as well as more minor ones such as minimal

Table 2.1: Literature Survey on the aircraft design aspects

Factors	Purpose	Influential parameters
Aerodynamic Considerations	To visualize the flow over the aircraft and aerodynamic characteristics	Shape, fineness ratio, AR, Lift–to–drag ratio
Performance & Operational Aspects	To investigate the flight maneuver capacity at a certain range, To adopt the prioritization degree of design	Maneuverability, thrust Fuel consumption, Speed, Take-off-landing distance, range, Altitude, etc
Structural Aspects	To analyze the structural integrity at various load factor	Payload capacity, MTOW Reliability, flight envelope
Socio-Economic Aspects	To identify the economical balance at different region and atmospheric condition	DOC, pricing, passengers comfort, atmospheric conditions & emission

human perception [69]. One of the most important aspects of the design phase is airflow over the aircraft [80]. The study of airflow over every type of element, regardless of size and shape, is referred to as aerodynamics [70]. The sizing and shaping are constantly changing both to include new information acquired about the design and to analyze the prospective design enhancements. Most airplanes are designed with streamlined profiles to reduce drag while increasing lift, resulting in reduced disturbance development on the airframe [46]. A wide wake pattern creates undesirable pressure gradients and causes an airplane to spin. This stalling phenomenon occurs regardless of flying speed or altitude. An unforeseen rise in stalling velocity causes a shift in the weight and CG balance allocation [70]. The propulsion unit has a relatively high level of influence on airflow assessment approaches and aircraft effectiveness [87]; it has a minor to moderate influence on environmental and economic problems [136]. Engine (or airframe) connections, fuel economy, noise and vibration, and so on are all aspects of aircraft propulsion technology [70]. Regardless of the passenger capacity [180], velocity and duration of flight equally influence the aircraft design. Due to the obvious rise in traveler payloads, a bigger engine is required. Additional research into the load transfer over the wing has strengthened it by 15 percent at a given flexural effectiveness. As a result, in enhancing propulsion systems, flight effectiveness must be improved. Aircraft designed to improve achieved by reducing load at a reasonable rate and distance of flight. During the touching down and turn phases, the flight design seems to have a significant effect on aviation results in terms of velocity, resistance, maneuvering, steadiness, as well as other factors. However, each aircraft can handle a variety of load factors, which may be investigated using the flying range. The influence of structural frame on flight data is significant and is only loosely related to aerodynamics considerations. Environmental variables such as life cycle [193], erosion, bird impact, rainstorms, wind, and so on may all have an impact on transport design [105]. The purpose of airplane structural design is to endure multiple kinds of air plus inertial stresses throughout various phases of flight. Reusable systems are being researched to extend their operational life [193]. According to Wesseler, the cost is determined by the airflow layout and propulsive systems [illegible] [15]. Enhanced fuel consumption is achieved by increasing the power and decreasing the resistance of aerodynamic attachments. However, it can increase effectiveness by extending the airplane's distance and decreasing its influence on the carbon footprint. As the proposed framework advances, judgments on the assessment of relevant designs, innovations, components, processes, architecture, approaches, and overall precisions should be made in Table 2.1. A figure of merits for every design feature must be calculated to certain other possibilities in market research. in-depth study of each critical design factor is studied and tabulated in Table 2.2.

2.1.1 Data Acquisition of aircraft

In the context of the manuscript's focus on wide-body aircraft design, the emphasis on these three design phases is essential. Starting with a well-defined conceptual phase al-

Table 2.2: Survey on the aircraft design significance

Selecting Parameters	Notations	Significance & Implementations	Reference(s)
Fuselage fuselage (m)	L_f	Enhances stability, fuel efficiency, and take-off rotation clearance.	[illegible] [illegible]
Fineness Ratio (no unit)	FR	drag estimation using a wetted area results in drag-reduction.	[illegible] [illegible]
Wing Area (m2)	Sw	enhancing lift coefficient, fuel efficiency, and increase structural flexibility.	[illegible] [illegible]
Aspect Ratio (no units)	AR	To determine the induced drag coefficient of the aircraft wing. Larger the AR triggers to decrease the vortices and hence to decrease drag.	[illegible] [illegible] [illegible] [illegible] [illegible]
Sweep Angle (degree)	Λ	To determine the induced drag coefficient and stall propagation, influencing aerodynamic performance.	[illegible] [illegible] [illegible] [illegible] [illegible]
Maximum Take-off Weight MTOW (kg)	W_{TO}	Signifies the decrease in fuel fraction by innovative technology can enhance long-range flight, better aerodynamic efficiency, and reduces TSFC in an aircraft. It strongly associated with payload, structures, speed, range, and fuel burn.	[illegible] [illegible] [illegible]
Empty weight Fraction (no units)	W_E/W_{TO}	It depends on the material selected and used in the structural airframes. Light weight can reduce emission and improve fuel efficiency.	(Proposed criteria)
Payload Weight Fraction (no units)	W_{PL}/W_{TO}	It determines the weight of the number of passengers, crews, and amount of cargo available in the flight.	[illegible]
Fuel Weight Fraction	W_F/W_{TO}	The amount of fuel required to cover a certain range of distance in a one load of fuel. The value can be determined by the mission profile	[illegible] [illegible] [illegible]
Payload	n_{PL}	The amount can be obtained by the number of passengers, crew, and cargo associated with competitiveness.	[illegible] [illegible] [illegible] [illegible]
Wing Loading (kg/m2)	W/S	Low W/S offer better sustained performance to create greater lift for the same amount of engine power. As a result, increasing W/S lengthens STO distances and reduces manoeuvrability.	[illegible] [illegible] [illegible] [illegible]
Thrust to Weight ratio (no units)	T/W	An aeroplane with a high T/W ratio will have a high thrust flying capabilities.	[illegible] [illegible]
Range (km)	R	The challenge of efficient range operating of an aircraft cover better distance out of a given fuel load. number of passengers, crew, and cargo associated with profits.	[illegible] [illegible] [illegible]
Take-off Landing Cycle (km)	STO	The main aim is to decrease the STO distance to improve the performance and fuel efficiency. By better design configuration can achieve improve STO.	[illegible] [illegible]

Cruise Speed (km/hr)	VC	Parameter is most economical to fly, somewhat quicker than the speed that allows for maximum range but slower than the aircraft's highest speed.	[ ][ ][ ][ ] [ ]
Altitude (ft)	ha	Impacts aircraft overall performance takeoff distance, velocity, climb rate, cruising speed, and fuel consumption.	[ ][ ][ ]
Rate of Climb (m/s)	ROC	Signifies the parameter of performance of the flight. A better ROC can be generated by greater the propulsion force, the lesser the resistive force, and the lowering the MTOW.	[ ]
Fuel Consumption (kg/hr/N)	SFC	In order to the proper selection of engine and aerodynamic configure the reduction of fuel consumption. Also by lowering the fuel use can lead lower the CO2 emissions.	[ ][ ][ ]
Thrust (kN)	T	Influences flight trajectory aircraft performance, and economics	[ ][ ]
Available Seat Kilometres (seat. km)	ASK	Capture the concept of demand and supply crucial for revenue, operational efficiency, optimizing seating and passenger experience.	[ ][ ]

lows for the integration of superior aerodynamics to reduce fuel burn and emissions, as well as to enhance passenger comfort [ ]. As the design progresses through the preliminary and detailed phases, the specific aspects of aerodynamics, structural engineering, and passenger cabin layout can be fine-tuned to optimize the overall performance and profitability of the wide-body aircraft. Current commercial trends in aircraft technologies such as Blended wing body [ ], double-bubble [ ], and V-flying aircraft [ ] are among the design principles currently considered as highly potential for future industry partners. Studies have noticed that among several transport aircraft, MoM with wide-body configuration generates substantial growth in profitability and fuel efficiency [ ] [ ]. The manuscript focused on wide-body fuselage design, and as part of the research, it collected a list of aircraft data to find correlation factors and identify models for performance improvement. This manuscript aims to study technical aspects and socio-economic aspects such as aerodynamic consideration, propulsive systems, operational performance, structural integrity, and profitability. These factors are illustrated and tabulated in Table 2.3, 2.4, 2.5, 2.6, 2.7. As reported by different authors [ ] which influences better revenue and fuel economy for the aviation sector. Emerging innovations are likely to result in even more significant fuel savings for passenger and freight aircraft in the coming years. In Chapter 1, this research work extensively elucidates the central theme of wide-body aircraft, offering a comprehensive analysis of both the prevailing market demands and the anticipated future trends in the aviation industry.

Table 2.3: Aerodynamic Consideration of transport aircraft

Manufacturer	Aerodynamic consideration					
	Length (m) (m)	Finess Ratio	Area (m^2) (m^2)	Span (m) (m)	Aspect Ratio	1/4 Chord Sweep (°)
AIRBUS A300-600	53.300	9.450	260.000	44.840	7.733	28.000
AIRBUS A310-200	46.660	8.270	219.000	43.900	8.800	28.000
AIRBUS A330-900	63.660	12.100	191.000	64.000	8.500	28.000
AIRBUS A330-300	62.470	11.076	363.100	58.000	9.265	29.700
AIRBUS A340-300	62.470	11.076	363.100	58.000	9.265	29.700
AIRBUS A340-600	69.570	12.335	437.300	61.200	8.565	31.100
AIRBUS A380-800	72.720	10.180	845.000	79.800	7.530	33.500
AIRBUS A350-900	66.800	11.910	442.000	64.750	9.490	31.900
AIRBUS A350-900	73.790	12.110	464.300	64.75	9.030	31.900
BOEING 747-100	68.630	10.558	511.000	59.640	6.961	37.500
BOEING 747-200	68.630	10.558	511.000	59.640	6.961	37.500
BOEING 747- 400	68.630	10.558	525.000	62.300	7.393	37.500
BOEING 777-200	62.780	10.126	427.800	60.900	8.669	31.600
BOEING 777-300	72.880	11.755	427.800	60.900	8.669	31.600
BOEING 787-9X	68.280	11.830	377.000	60.120	9.590	32.200
BOEING 777-9X	76.730	12.380	516.700	71.750	9.960	31.640
DOUG. DC10-30	51.970	8.633	367.700	50.400	6.908	35.000
McDON. MD - 11	58.650	9.743	338.900	51.770	7.908	35.000
LOCKHDL1011-100	54.200	8.944	321.500	47.350	6.974	35.000
ILYUSHIN II-86	56.100	9.227	320.000	48.060	7.218	35.000
ILYUSHIN II-96-300	51.150	8.413	391.600	55.570	7.886	30.000
ILYUSHIN II-96	60.500	9.951	391.600	55.570	7.886	30.000
AIRBUS A319-100	33.840	8.567	122.400	33.910	9.395	25.000
AIRBUS A320-200	37.570	9.511	122.400	33.910	9.395	25.000
AIRBUS A321-200	44.510	11.268	122.400	33.910	9.395	25.000
BOEING B707-320	44.350	7.300	283.400	44.420	6.962	35.000
BOEING B727-200	41.510	7.000	157.900	32.920	6.863	32.000
BOEING B737-200	29.540	7.400	91.040	28.350	8.828	25.000
BOEING B737-300	32.300	7.400	91.040	28.900	9.174	25.000
BOEING B737-400	35.300	7.400	91.040	28.900	9.174	25.000
BOEING B737-500	29.900	7.400	91.040	28.900	9.174	25.000
BOEING B737-600	29.880	7.400	124.600	34.300	9.442	25.000
BOEING B737-700	32.180	7.400	124.600	34.300	9.442	25.000
BOEING B737-800	38.080	7.400	124.600	34.300	9.442	25.000
BOEING B757-300	46.960	11.740	185.250	38.050	7.815	25.000
BOEING B757-300	53.960	13.490	185.250	38.050	7.815	25.000
BOEINGB767-200	47.240	9.392	283.300	47.570	7.988	31.500
BOEINGB767-300	53.670	10.670	283.300	47.570	7.988	31.500
BOEINGB767-300 ER	53.670	10.670	283.300	47.570	7.988	31.500
DOUG.DC 8-63	55.700	14.658	271.900	45.230	7.524	30.000
DOUG.DC 9 - 50	37.140	10.317	92.970	28.470	8.718	24.000
McDON. MD-81	41.300	11.440	112.300	32.870	9.621	24.500

Table 2.4: Propulsive systems of transport aircraft

Manufacturer	Propulsive system Engine Manufacturer	No. of engines	Endurance	Total thrust (kN)
AIRBUS A300-600	PW	2	8.361	514.000
AIRBUS A310-200	PW	2	7.647	514.800
AIRBUS A330-900	PW	2	13.268	648.000
AIRBUS A330-300	GE	2	13.517	600.000
AIRBUS A340-300	CFMI	4	15.005	604.000
AIRBUS A340-600	R-R	4	15.956	996.400
AIRBUS A380-800	R-R	4	14.013	1392.000
AIRBUS A350-900	R-R	2	15.873	749.000
AIRBUS A350-900	R-R	2	17.037	863.000
BOEING 747-100	PW	4	10.437	835.200
BOEING 747-200	R-R	4	13.106	944.000
BOEING 747- 400	PW	4	14.416	1009.600
BOEING 777-200	R-R/GE/PW	2	14.737	747.320
BOEING 777-300	R-R	2	11.753	846.000
BOEING 787-9X	R-R	2	12.317	680.000
BOEING 777-9X	GE	2	19.687	934.000
DOUG. DC10-30	GE	3	9.811	708.000
McDON. MD - 11	GE	3	13.952	822.000
LOCKHD L1011-100	R-R	3	6.967	561.000
ILYUSHIN II-86	Kuznetsov	4	11.507	510.000
ILYUSHIN II-96-300	Soloviev	4	13.529	627.600
ILYUSHIN II-96	PW	4	11.494	658.400
AIRBUS A319-100	CFMI	2	14.574	199.400
AIRBUS A320-200	CFMI	2	7.367	222.400
AIRBUS A321-200	CFMI	2	7.111	284.000
BOEING B707-320	PW	4	10.460	338.160
BOEING B727-200	PW	3	5.760	213.600
BOEING B737-200	PW	2	6.299	142.400
BOEING B737-300	CFMI	2	5.286	178.000
BOEING B737-400	CFMI	2	4.897	195.800
BOEING B737-500	CFMI	2	6.923	164.600
BOEING B737-600	CFMI	2	7.199	164.000
BOEING B737-700	CFMI	2	7.622	178.000
BOEING B737-800	CFMI	2	6.809	214.000
BOEING B757-300	R-R	2	8.529	382.600
BOEING B757-300	R-R	2	7.520	384.000
BOEINGB767-200	GE	2	14.358	427.000
BOEINGB767-300	GE	2	13.029	447.000
BOEINGB767-300 ER	GE	2	7.058	547.200
DOUG. DC 8-63	PW	4	8.279	320.400
DOUG. DC 9 - 50	PW	2	2.948	137.800
McDON. MD-81	PW	2	4.101	171.200

Table 2.5: Structural Aspects of transport aircraft

Manufacturer	Structural Aspects			
	MTOW	WE/WTO	WPL/WTO	Max. seats
AIRBUS A300-600	170500	0.521	0.241	375
AIRBUS A310-200	144000	0.551	0.228	275
AIRBUS A330-900	227703	0.550	0.221	460
AIRBUS A330-300	217000	0.545	0.223	440
AIRBUS A340-300	271000	0.479	0.178	440
AIRBUS A340-600	365000	0.485	0.173	475
AIRBUS A380-800	575000	0.482	0.146	853
AIRBUS A350-900	280000.3	0.413	0.191	440
AIRBUS A350-900	286670	0.491	0.215	480
BOEING 747-100	340195	0.497	0.205	516
BOEING 747-200	377840	0.466	0.166	516
BOEING 747- 400	396830	0.457	0.154	660
BOEING 777-200	286897	0.481	0.198	440
BOEING 777-300	299370	0.521	0.229	550
BOEING 787-9X	254011	0.533	0.225	440
BOEING 777-9X	352441.3	0.511	0.208	426
DOUG. DC10-30	263636	0.460	0.183	380
McDON. MD - 11	283720	0.473	0.196	405
LOCKHD L1011-100	211374	0.529	0.158	400
ILYUSHIN II-86	208000	0.400	0.202	350
ILYUSHIN II-96-300	216000	0.542	0.185	300
ILYUSHIN II-96	270000	0.490	0.215	375
AIRBUS A319-100	64000	0.613	0.272	153
AIRBUS A320-200	73500	0.562	0.261	179
AIRBUS A321-200	89000	0.539	0.256	220
BOEING B707-320	151315	0.438	0.252	219
BOEING B727-200	95028	0.486	0.196	189
BOEING B737-200	52390	0.528	0.295	130
BOEING B737-300	56470	0.564	0.284	149
BOEING B737-400	62820	0.531	0.282	170
BOEING B737-500	52390	0.591	0.296	130
BOEING B737-600	65090	0.560	0.151	132
BOEING B737-700	69400	0.542	0.167	149
BOEING B737-800	78220	0.530	0.188	189
BOEING B757-300	115900	0.501	0.222	239
BOEING B757-300	122470	0.539	0.241	289
BOEINGB767-200	136078	0.595	0.250	255
BOEINGB767-300	156489	0.557	0.250	290
BOEINGB767-300 ER	181437	0.495	0.251	290
DOUG. DC 8-63	158760	0.439	0.193	259
DOUG. DC 9 - 50	54420	0.538	0.281	139
McDON. MD-81	63500	0.560	0.283	172

Table 2.6: Operational performance of transport aircraft

Manufacturer	Operational performance						
	W/S (kg/m2)	T/W	LTO	Vc (km/hr)	R (km)	ha (ft)	RoC m/s
AIRBUS A300-600	655.769	0.307	2280	897.0	7500	31000	16.26
AIRBUS A310-200	655.770	0.307	2280	850.0	6500	41100	16.26
AIRBUS A330-900	657.530	0.307	2280	1005.0	13334	33000	16.26
AIRBUS A330-300	597.632	0.282	2770	870.0	11760	33000	10.16
AIRBUS A340-300	746.351	0.227	3000	913.0	13700	33000	28.96
AIRBUS A340-600	834.667	0.278	3140	915.0	14600	41450	28.96
AIRBUS A380-800	680.470	0.247	3000	1099.0	15400	43000	7.62
AIRBUS A350-900	633.480	0.273	2600	945.0	15000	43100	10.16
AIRBUS A350-900	633.480	0.271	2600	945.0	16100	41450	10.16
BOEING 747-100	665.744	0.250	3246	939.0	9800	35000	5.08
BOEING 747-200	739.413	0.255	3200	969.0	12700	45100	5.08
BOEING 747- 400	755.867	0.259	3310	933.0	13450	35000	7.62
BOEING 777-200	670.633	0.266	2438	950.0	14000	43100	15.24
BOEING 777-300	699.790	0.288	3230	950.0	11165	43100	15.24
BOEING 787-9X	673.700	0.273	2800	954.0	11750	41100	15.24
BOEING 777-9X	682.100	0.270	3000	1034.8	20372	43100	15.24
DOUG. DC10-30	716.987	0.274	3000	981.6	9630	42000	10.16
McDON. MD - 11	837.179	0.295	3139	961.0	13408	43000	10.16
LOCKHD L1011-100	657.462	0.271	2350	957.0	6667	33000	14.22
ILYUSHIN II-86	650.000	0.250	2700	869.0	10000	43000	15.00
ILYUSHIN II-96-300	650.000	0.250	2400	850.0	11500	43000	15.00
ILYUSHIN II-96	650.000	0.250	2600	870.0	10000	43000	15.00
AIRBUS A319-100	522.876	0.318	1950	826.0	12038	33000	15.24
AIRBUS A320-200	600.490	0.308	2180	829.7	6112	28000	12.70
AIRBUS A321-200	727.124	0.325	2000	833.4	5926	28000	12.70
BOEING B707-320	533.927	0.228	35550	885.3	9260	25000	18.03
BOEING B727-200	601.824	0.229	3033	864.9	4982	25000	12.70
BOEING B737-200	575.461	0.277	1829	777.8	4900	25000	7.62
BOEING B737-300	620.277	0.321	2300	794.5	4200	26000	12.70
BOEING B737-400	690.026	0.318	2540	796.4	3900	26000	15.24
BOEING B737-500	575.461	0.320	2650	794.5	5500	26000	15.24
BOEING B737-600	522.392	0.257	1900	833.4	6000	41000	15.24
BOEING B737-700	556.982	0.261	2000	837.1	6380	41000	15.24
BOEING B737-800	627.769	0.279	2316	837.1	5700	41000	15.24
BOEING B757-300	625.641	0.337	2226	850.1	7250	31000	17.78
BOEING B757-300	661.107	0.320	2600	853.8	6420	31000	17.78
BOEINGB767-200	480.332	0.320	1770	850.1	12205	39000	10.16
BOEINGB767-300	552.379	0.291	2550	851.9	11100	39000	15.24
BOEINGB767-300 ER	640.441	0.307	2550	850.1	6000	39000	15.24
DOUG. DC 8-63	583.891	0.217	3505	895.0	7410	27000	10.16
DOUG. DC 9 - 50	585.350	0.258	2100	816.7	2408	27000	10.16
McDON. MD-81	565.450	0.275	2195	813.0	3334	27000	15.24

Table 2.7: Socio-economical factors of transport aircraft

Manufacturer	Socio-economic competitiveness Seat *range	SFC
AIRBUS A300-600	2812500	0.01253
AIRBUS A310-200	1787500	0.01218
AIRBUS A330-900	6133640	0.01857
AIRBUS A330-300	5174400	0.00957
AIRBUS A340-300	6028000	0.01233
AIRBUS A340-600	6935000	0.00972
AIRBUS A380-800	13136200	0.01303
AIRBUS A350-900	6600000	0.00928
AIRBUS A350-900	7728000	0.00751
BOEING 747-100	5056800	0.01798
BOEING 747-200	6553200	0.01305
BOEING 747- 400	8877000	0.01109
BOEING 777-200	6160000	0.01214
BOEING 777-300	6140750	0.01360
BOEING 787-9X	5170000	0.01210
BOEING 777-9X	8678472	0.00864
DOUG. DC10-30	3659400	0.01571
McDON. MD - 11	5430240	0.01048
LOCKHD L1011-100	2666800	0.02021
ILYUSHIN II-86	3500000	0.01535
ILYUSHIN II-96-300	3450000	0.01399
ILYUSHIN II-96	3750000	0.01570
AIRBUS A319-100	1841814	0.00649
AIRBUS A320-200	1094048	0.01151
AIRBUS A321-200	1303720	0.00927
BOEING B707-320	2027940	0.02017
BOEING B727-200	941598	0.01966
BOEING B737-200	637000	0.01990
BOEING B737-300	625800	0.01688
BOEING B737-400	663000	0.01656
BOEING B737-500	715000	0.01394
BOEING B737-600	792000	0.01741
BOEING B737-700	950620	0.01515
BOEING B737-800	1077300	0.01411
BOEING B757-300	1732750	0.01031
BOEING B757-300	1855380	0.01190
BOEINGB767-200	3112275	0.00815
BOEINGB767-300	3219000	0.00857
BOEINGB767-300 ER	1740000	0.01869
DOUG. DC 8-63	1919190	0.02736
DOUG. DC 9 - 50	334712	0.03134
McDON. MD-81	573448	0.02488

Table 2.8: Aerodynamic consideration of midsize aircraft

Manufacturer	Aerodynamic Configuration							
	Length	Cabin width	Fuselage width	Fuselage height	Aisle type	Fineness ratio	Aspect ratio	Aerodynamic accessories
	m	m	m	m	-	-	-	Winglets, flaps, etc.
Airbus A319-100	33.84	3.70	3.95	4.14	Single aisle	8.57	9.39	Slats, sharklets, flaps
Airbus A320-200	37.57	3.70	3.95	4.14	Single aisle	9.51	9.39	sharklets, flaps
Airbus A321-200	44.51	3.70	3.95	4.14	Single aisle	11.27	9.39	sharklets, flaps
Boeing B737-500	37	3.54	3.80	-	Single aisle	7.40	9.17	blended winglet
Boeing B737-600	42	3.54	3.80	-	Single aisle	7.40	9.44	
Boeing B737-700	42	3.54	3.80	-	Single aisle	7.40	9.44	
Boeing B737-800	42	3.54	3.80	-	Single aisle	7.40	9.44	
Boeing B757-300	47.3	3.54	3.76	5.41	Single aisle	13.49	7.82	
Boeing B767-200	48.51	4.72	5.03	5.41	Single aisle	9.39	7.99	
Airbus A310-200	46.66	5.28	5.64	4.33	Twin-aisle	8.27	8.80	
Airbus A330-300	63.67	5.26	5.64	4.33	Twin-aisle	11.08	9.26	
Boeing B777-300	73.86	5.86	6.20	-	Twin-aisle	11.75	8.67	
Boeing B787-10	68.28	5.49	5.77	5.94	Twin-aisle	11.83	9.59	Raked Winglets

2.2 Comparative Study

2.2.1 Structural Integrity

Airframe structural integrity provides better performance and fuel economy possible through the mechanical aspects of aircraft structures (Hong et al., 2020). Aircraft structural design parameters reflect the loads acting on an aircraft, material used, structural weight, and weight ratios. Table 2.9 shows the structural mass fraction of both aircraft variants.

Table 2.9: Structural mass of Wide-bodies and Narrow-bodies aircraft

Weights	**Notation**	**Wide-body aircraft**	**Narrow-body aircraft**
No. of passenger	npax	200 – 853	130 – 290
MTOW	W_{TO} (kg) * 103	144 – 575	52 – 181
Payload Weight	W_{PL} (kg)	32800 – 84000	9800 – 45600
Volume per seat	(m^3)	0.20 – 0.54	0.18 – 0.34
Empty weight fraction	WE/WO	0.4 – 0.55	0.44 – 0.61
Fuel Weight fraction	WFuel/WO	0.32 – 0.7	0.21 – 0.47

The supreme difference in both variants is the number of passengers (npax) capacity. Seats add a significant amount of MTOW. Wide-body aircraft are capable of carrying numerous passengers in a single fleet. Emerging to increase in maximum take-off weight (W_{TO}) includes operating empty weight (W_E), Payload weight (W_{PL}), and fuel weight (W_{Fuel}), mathematically expressed in the equation (2.2.1) and 2.2.2. The allocation of cabin crew is contingent upon the number of passengers, thereby determining the requisite provision of lavatories and galleys within the aircraft's cabin. As per FAR, the weight of each passenger is limited to 119 kg which includes body weight and passenger luggage weight. Another significant parameter that interprets materials and structural weight is the empty weight fraction (W_E/W_{TO}). W_E/W_{TO} consists of structural, avionics, and systems weights expressed in equation (2.2.3). Investigating the aircraft variants noticed that W_E/W_{TO} limited the value from 0.4 to 0.6 following the standard thumb rule as shown in Figure. Furthermore, a detailed expansion form of the empty weight is mathematically expressed in the equation (2.2.4). This equation 2.2.4 clearly states that depends on configuration, the material used, and the component installation location in Table 2.10.

$$W_{TO} = W_{PL} + W_{CARGO} + W_{Fuel} + W_E \tag{2.2.1}$$

$$W_{TO} = \frac{W_{PL} + W_{CARGO} + W_E}{\left(1 - \frac{W_{Fuel}}{W_{TO}}\right)} \tag{2.2.2}$$

$$W_E = W_W + W_F + W_{HT} + W_{VT} + W_{LG} + W_{Eng} + W_{avionics} \tag{2.2.3}$$

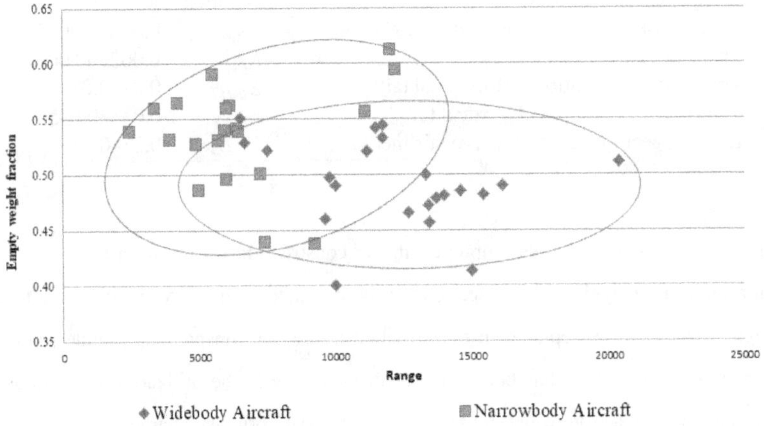

Figure 2.2: Variation of empty weight fraction for different aircraft configuration

$$W_W = S_W C_{mac} \left(\frac{t}{c}\right)_{max,W} \rho_{mat,W} K_{\rho,W} \left(\frac{ARn_{ult}}{cos\Lambda_{0.25c}}\right)^{0.6} \lambda^{0.04} g \tag{2.2.4a}$$

$$W_F = L_f D_f^2 \rho_{mat,f} K_{\rho,f} n_{ult}^{0.25} K_{inlet} g \tag{2.2.4b}$$

$$W_{HT} = S_{HT} C_{mac,HT} \left(\frac{t}{c}\right)_{max,HT} \rho_{mat,HT} K_{\rho,HT} \left(\frac{AR_{HT}}{cos\Lambda_{0.25c,HT}}\right)^{0.6} \lambda_{HT}^{0.04} \overline{V_{HT}}^{-0.3} \left(\frac{C_e}{C_{HT}}\right)^{0.4} g \tag{2.2.4c}$$

$$W_{VT} = S_{VT} C_{mac,VT} \left(\frac{t}{c}\right)_{max,VT} \rho_{mat,VT} K_{\rho,VT} \left(\frac{AR_{VT}}{cos\Lambda_{0.25c,VT}}\right)^{0.6} \lambda_{VT}^{0.04} \overline{V_{VT}}^{-0.2} \left(\frac{C_r}{C_{VT}}\right)^{0.4} g \tag{2.2.4d}$$

$$W_{LG} = K_L K_{ret} K_{\rho,LG} W_L \left(\frac{h_{LG}}{b_W}\right) n_{ult}^{0.2} g \tag{2.2.4e}$$

$$W_{Eng,ins} = K_E N_E W_{Eng}^{0.9} \tag{2.2.4f}$$

$$\frac{W_{avionics}}{W_{empty}} = 0.02 \tag{2.2.4g}$$

Table 2.10: Component density factor of transport aircraft

Component Installation	Notation	K_ρ
Engine installed on the wing, fuel tank in the wing	$K_{\rho,W}$	0.003–0.0035
Transport, cargo, airliner fuselage	$K_{\rho,f}$	0.0025–0.0032
Transport – conventional Horizontal tail	$K_{\rho,HT}$	0.02 0.03
Transport – conventional Vertical tail	$K_{\rho,VT}$	0.035–0.045
Landing gear of transport, cargo, airliner	$K_{\rho,LG}$	0.28–0.35

The proportion of materials employed in the construction of an aircraft plays a pivotal role, bearing significant implications for both structural integrity and overall aircraft weight. The data, as depicted in the provided figure 2.3 conspicuously underscores a noteworthy trend - there has been a noticeable increase in the utilization of composite materials in comparison to previous periods [126]. The primary characteristic that renders composite materials particularly valuable in this context is their inherent capacity to achieve a dual objective: reducing the overall weight of the aircraft while concurrently enhancing its stiffness. Notably, in the realm of contemporary aircraft design, the Airbus A350 series stands out as an exemplar of technological advancement. This aircraft

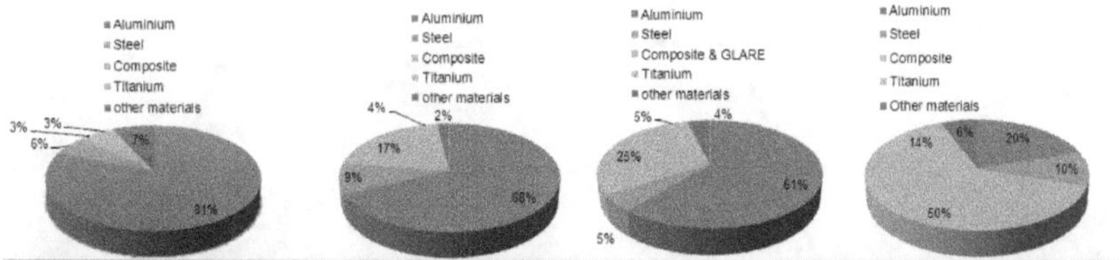

Boeing 737	Airbus 340 - 330	Airbus A 380	Boeing 787
Aluminium - 81 %	Aluminium - 68 %	Aluminium - 61 %	Aluminium - 20 %
Steel - 6%	Steel - 9%	Steel - 9%	Steel - 10%
Composite - 3%	Composite - 17%	Composite - 17%	Composite - 50%
Titanium - 3%	Titanium - 4%	Titanium - 5%	Titanium - 14%
Other materials - 7%	Other materials - 2%	Other materials - 4%	Other materials - 6%

Figure 2.3: Percent of material used transport aircraft

series boasts the most cutting-edge iteration, characterized by an impressively low empty weight fraction, specifically quantified at 0.413. This achievement represents a significant milestone in the ongoing pursuit of maximizing the efficiency and performance of modern aircraft.

Considering the third most important factor, the fuel weight fraction (W_{Fuel}/W_{TO}) value is different for both aircraft variants and completely depends on the flight mission as shown in Figure 2.4. Figure 2.4 represents the entire mission sequence, commencing from the initial stage labeled "Engine Warm-Up (1)" and progressing through various stages until reaching the 8th stage marked as "Landing and Engine Shut-Off (8)." These stages encapsulate the entirety of the mission from engine startup to shutoff. In addition to it, a small amount of fuel is kept as trapped by the systems and reserved for emergency purposes about 6 to 8 percent of the total fuel [14] [15] is expended throughout these mission stages is quantitatively represented by a specific equation 2.2.5. This equation (2.2.5) is assumed as the fuel reserved and trapped about 6 percent of fuel.

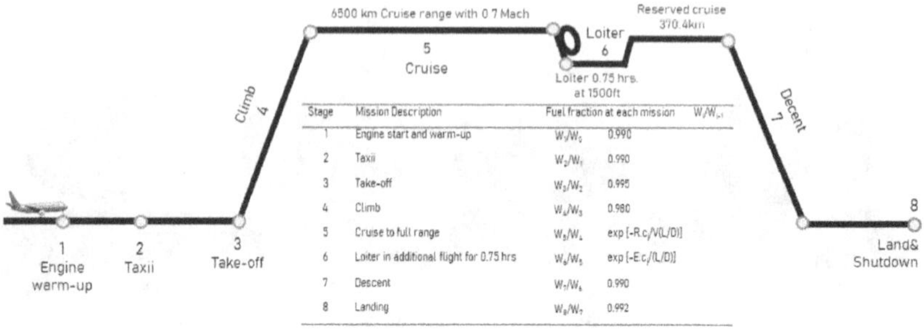

Figure 2.4: Flight mission of transport aircraft

$$\frac{W_{Fuel}}{W_{TO}} = 1.06\left(1 - \frac{W_8}{W_{TO}}\right) \tag{2.2.5}$$

where,

$$\frac{W_8}{W_{TO}} = \frac{W_8}{W_7} * \frac{W_7}{W_6} * \frac{W_6}{W_5} * \frac{W_5}{W_4} * \frac{W_4}{W_3} * \frac{W_3}{W_2} * \frac{W_2}{W_1} * \frac{W_1}{W_{TO}}$$

This equation 2.2.5 serves as a comprehensive model that tracks and calculates fuel consumption throughout the entire mission profile from W1 to W8. It takes into account various factors, such as engine performance, aircraft weight, altitude, speed, and other relevant parameters to estimate how fuel is utilized at each stage of the mission. By utilizing this equation, it becomes possible to precisely understand and predict fuel consumption patterns throughout the mission, which is crucial for optimizing aircraft performance, range, and efficiency. The Breguet equation (2.2.6) represents that fuel fraction is based on the range of the flight, endurance, flight speed, engine type, specific fuel consumption, and lift-to-drag ratio. This equation signifies the decrease in fuel fraction by innovative technology can enhance long-range flight, better aerodynamic efficiency, and reduce TSFC in an aircraft. These factors are quantitative influence was found to be very sensitive to airframe configuration. On average, fuel weight for wide-body aircraft requires 50 percent of MTOW whereas, narrow-body aircraft needs 34 percent of MTOW, which is more compared to narrow-body aircraft.

$$R = \frac{L}{D}\frac{V_c}{C_T}\ln\frac{W_{fuel}}{W_{TO}} \tag{2.2.6}$$

Where R is Range; $\frac{L}{D}$ is aerodynamic efficiency; V is Flight velocity; C_T is TSFC; $\frac{W_{fuel}}{W_{TO}}$ is fuel weight fraction.

2.2.2 Aerodynamic Configurations

Based on the literature review, fuel economy can be improved by better aerodynamic configuration. Potential changes in wingspan and fuselage length are sensitive to fuel economy noticed from Table 2.11. Constructing Table 2.11 illustrates an aircraft's aerodynamic configuration like Fuselage length(L_f), fineness ratio (FR), abreast, aisle, Wing Span (b_W), Wing Aspect Ratio (AR_W), Taper Ratio (λ_W), Wing Sweep angle ($\Lambda_{LE,W}$), average thickness–to-chord ratio (t/c), high lifting device (HLD), VT height (h_{HT}), VT tail lever arm (Lt_{VT}), VT Aspect Ratio ($AR_{VT} = \frac{b_{VT}^2}{S_{VT}}$), HT Taper Ratio ($\lambda_{HT}$), VT Sweep angle ($\Lambda_{VT}$), HT Span ($b_{HT}$), HT tail lever arm ($Lt_{HT}$), HT Aspect Ratio($AR_{HT}$), HT Taper Ratio ($\Lambda_{HT}$), HT Sweep angle ($\Lambda_{HT}$), Track ($l_{LG}$), wheelbase ($b_{LG}$), nacelle length(ln), and nacelle span location are mathematically expressed in equation 2.2.7, 2.2.8, 2.2.9, 2.2.10, 2.2.11 and 2.2.12.

$$FR = \frac{L_f}{D_f} \tag{2.2.7}$$

$$AR_{W,HT,VT} = \frac{b_{W,HT,VT}^2}{S_{W,HT,VT}} \tag{2.2.8}$$

$$C_{R,W,HT,VT} = \frac{2.S_{W,HT,VT}}{b_{W,HT,VT}.\left(1+\lambda_{W,HT,VT}\right)} \tag{2.2.9}$$

$$\tan\Lambda_{LE,W} = \tan\left(\Lambda_{c/4}\right) - \left(\frac{S_W}{b_W^2}\frac{1-\lambda}{1+\lambda}\right) \tag{2.2.10}$$

$$V_W = \frac{S_W^2}{3b_W^2\left(1+\lambda_W^2\right)}\left[(K_R+3)\left(\frac{t}{c}\right)_R + (K_T+3)\left(\frac{t}{c}\right)_T \lambda_W^2\right] \tag{2.2.11}$$

$$\lambda_{W,HT,VT} = \frac{C_{T_{W,HT,VT}}}{C_{R_{W,HT,VT}}} \tag{2.2.12}$$

According to research, drag accounts for 75% of overall airplane fuel consumption. In general, an aircraft's wing and fuselage provide the greatest drag, and some studies have shown that the fuselage produces half or one-third of the overall drag as illustrated in Fig 2.5. Fuselage drag is determined by the material used in its construction as well as the fuselage shape. There is only one parameter that influences the drag according to the fuselage geometry is the fuselage fineness ratio which is the ratio of the length of the fuselage to the diameter of the fuselage. The main motto behind the design of wide-body aircraft is to combine efficiency and passenger comfort and to expand the amount of cargo space in a plane. The wide body fuselage gives a wide wing span that allows more efficient lift and stability during flight [[illegible]]. A wide body can be distinct from a narrow body in many ways. Usually, wide-body aircraft have twin-aisle, and narrow-body aircraft have a single aisle shown in Figure 2.6. A Wide-body configuration imparts a larger passenger density than the narrow-body. Based on the collected case study, wide-body fuselage length is longer than narrow-body aircraft. Unlike narrow-body, the FR is limited to the range 8 - 12 for wide-body aircraft. The FR is a significant parameter to determine the drag estimation using a wetted area as expressed in equation (2.2.13). The lower the value of FR, the lower the value of the wetted area resulting in drag-reduction[[illegible]] [[illegible]].

$$C_{DO_{component}} = C_{DO_{wing}} + C_{DO_{Fuselage}} + C_{DO_{HT}} + C_{DO_{VT}} + C_{DO_{nacelle}} \tag{2.2.13}$$

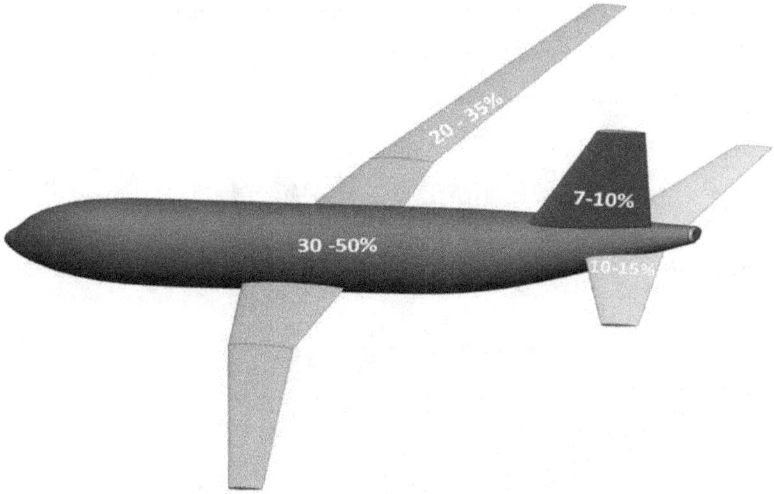

Figure 2.5: Drag contribution by the transport aircraft

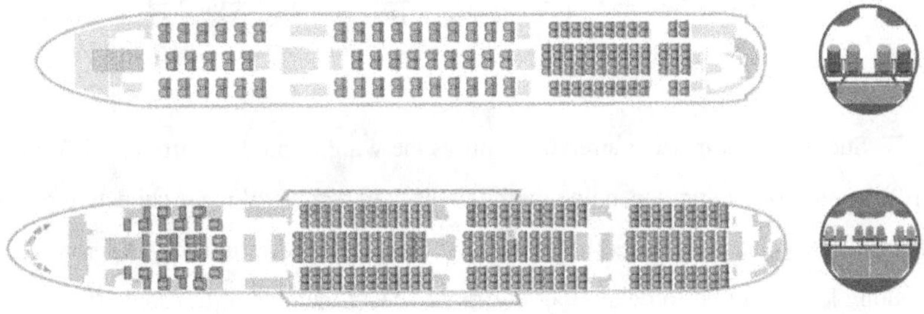

Figure 2.6: Single-aisle and Twin-aisle fuselage configuration

In another form, the equation 2.2.13 can also be written as

$$C_{DO_{Component}} = \sum_{K=1}^{Components} (C_{f,e}).FF.Q.(S_{wet}/S_{ref}) \tag{2.2.14}$$

Where C_f is the Skin friction drag coefficient depending on Reynolds number, skin roughness, fuselage length, and speed. F.F Form factor depends on the fineness ratio; Q is the Interference factor depends on location.

$$C_{f,e} = \frac{0.455}{(lgRe)^{2.58}(1+0.144M^2)^{0.65}} \tag{2.2.15}$$

$$FF_{W,Ht,VT} = \left[1+\frac{0.6}{(x/c)_m}\left(\frac{t}{c}\right)+100\left(\frac{t}{c}\right)\right]\left[1.34M^{0.18}\left(\cos\Lambda_{c/4}\right)^{0.28}\right] \tag{2.2.16}$$

$$FF_f = \left[1+\frac{60}{(FR)^3}+\frac{FR}{400}\right] \tag{2.2.17}$$

$$S_{wet,w} = 4*\left(\frac{b}{2}\right)_e\left(\frac{C_R,e+C_T}{2}\right)\left(1+0.25\frac{t}{c}\right) \tag{2.2.18}$$

$$S_{wet,f} = \pi.L_f.D_f.\left(1-\frac{2}{FR}\right)^{2/3}\left(1+\frac{1}{FR^2}\right) \tag{2.2.19}$$

$$S_{wet,HT,VT} = 4*\left(\frac{b}{2}\right)_e\left(\frac{C_{R,HT,e}+C_{T,HT}}{2}\right)\left(1+0.25\frac{t}{c}\right) \tag{2.2.20}$$

Succeeding important aircraft feature is the wing to produce lift. An airfoil is the cross-section of a wing depending on the flight flow regime. Wing architecture changes depending on the flight speed affect the thickness-to-chord ratio (t/c). Appropriate selection of t/c required to delay drag divergence Mach number at transonic flight speed. A longer wingspan (bw) was noticed for long-range flight in wide-body compared to narrow-body. Mostly aerodynamic wingspan seems to be longer for wide-body aircraft for long-range and lowers the drag. Negligible changes are noticed in both aircraft's wing Aspect ratio(A.Rw), taper Ratio (λ_W), and wing Sweep angle (Λ_W). Sweep wing creates thinner t/c leading to delayed shock formation over the wing. Recent wing design configured with winglet initiates to reduce induced drag. New design pattern configured

Table 2.11: Aerodynamic configurations of Wide-bodies and Narrow-bodies aircraft

Components	Configuration	WB aircraft	NB aircraft
Fuselage	Lf (metre)	46 – 77	29 – 56
	F.R	8.2 – 12.4	4.3 – 14.7
	$N_{Abreast}$	7 – 10	5 – 8
	Aisle	2	1
Wing	b$_w$(*metres*)	43.5 – 79.8	28.3 – 47.6
	A.Rw	6.9 – 9.9	6.8 – 9.6
	λ_W	0.15 – 0.39	0.18 – 0.31
	Λ_W (degree)	28.0 – 37.5	24 – 35
	Average t/c (%)	9 – 12	9 – 15
HLD	Trailing Edge*	F2,S1, S2	S1, S2, S3
	Leading Edge	Slats, Kruger	Slats, flaps
	Flap Span/Wing Span	0.62 – 0.80	0.59 – 0.78
Vertical Tail	h_{VT} (metres)	8 – 11.2	3.9 – 9.0
	lt_{VT} (% of Lf)	35.5 – 46.7	31.5 – 41.4
	AR_{VT}	1.1 – 2.2	0.64 – 3.18
	λ_V	0.28 – 0.48	0.29 – 0.78
	Λ_{VT} (degree)	35 – 46	30 – 53
Horizontal Tail	b_{HT} (metres)	16.2 – 22.1	10.8 – 18.6
	LtHT (percent of Lf)	32.5 – 50.5	34.5 – 51.5
	A.RHT	3.5 – 5.0	3.3 – 5.6
	λ_{HT}	0.26 – 0.42	0.19 – 0.4
	Λ_{HT} (degree)	30.0 – 37.5	27.5 – 36.0
Undercarriage	lLG (metre)	9.6 – 11.2	4.8 – 11.00
	bLG (metres)	18.6 – 32.5	11.0 – 23.6
	No. of wheels (nose; main)	2;8 – 2;16	2;4 – 2;8
Nacelle	ln (metres)	4.9 – 7.3	4.4 – 7.0
	Span location (% of Lf)	30 – 36	28 – 35

with variable camber wing, strakes, and riblets. Additional wing fairings establish stability and avoid turbulency. Apart from the wing, HLD was installed to develop additional increment in lift either by leading edge or trailing edge devices like slats, Kruger, flaps, single-slotted, and double-slotted [105]. Currently, operational wide-body aircraft prefer to use double-slotted Fowler flaps, whereas narrow-body prefer to use triple-slotted flaps.

$$\left(\frac{C_L}{C_D}\right)_{max} = \sqrt{\frac{2\left(\frac{W}{S}\right)}{\rho\sqrt{\frac{C_{D,0}}{k}}}} \tag{2.2.21}$$

Looking into VT and HT, conventional tailplane configurations are the topmost priority

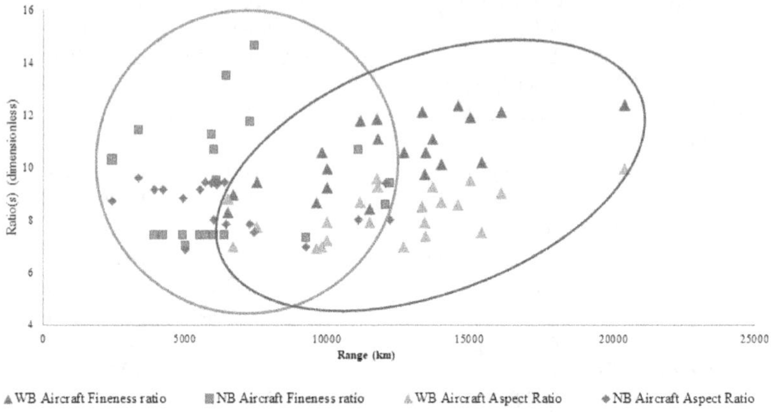

Figure 2.7: FR and AR for different aircraft configuration

for all manufacturers like Boeing, Airbus, Mc Douglas, and many more. Tailplane dimensioning for wide-body and narrow-body aircraft does not much difference. A tailplane is meant to improve stability depending on the size of the flight[84]. The location of the tail lever arm justifies the empennage drag and stability as well. In addition, the type of landing gear used in both aircraft is the same as multi-bogey but, the number of wheels used in aircraft is different from each other. Another important aerodynamic feature is the nacelle length and its location. These features do not affect the type of aircraft. Presently the aircraft manufacturers urge on advanced technology, the safety of passengers, maintaining sustainable flight, and lowering the operating cost. Business expenses are maintained lower in such a global marketplace. Enhanced fuel economy not only saves the economy of the airline industry but minimizes the overall aviation industry's carbon footprint [ ] [ ]. Fuel efficiency improved by better design and thrust output. Results to enhance the aircraft performance and propulsive systems.

2.2.3 Operational and Propulsive Aspects

Performance of the aircraft can be measured using cruise speed of aircraft (Vc)max, range (R), endurance (E), service Ceiling, Rate of climb (ROC), Wing loading (W_{TO}/S), Take-off Field Length (S_{TO}), and available seat kilometer (ASK) [117] [155] [21] [97] [158]. Table 2.12 represents the performance and propulsive systems of both aircraft. As per the data collection, wide-body cruise speed and range of flight are higher than narrow-body aircraft. Due to that, endurance is slightly higher in wide-body aircraft for long flights. ASK is a benchmark that defines the loss or profit for the airlines and is mathematically expressed in an equation. As a result, ASK seems to be more than narrow-body aircraft revealing an increase in passenger-providing hike in air transportation needs [117]. According to the IATA Feb 2023: Air Passenger Market Analysis report, the Asia-Pacific region shows a maximum increase in the difference in available seat kilometers (ASK) for international air traffic by about 153.3%. Efficient air traffic management decreases interruptions and enhances airline functionality. The core idea is that during a flight, various factors like altitude, speed, weight, weather conditions, aircraft configuration, route optimization, engine efficiency, and air traffic management interact with one another and with the specific phase of the mission to collectively impact the overall performance of the flight. Understanding these mathematical relationships is essential for optimizing performance in each phase and ensuring a successful and efficient flight mission as equation 2.2.23, 2.2.24, and 2.2.25. In addition, service ceiling, ROC, Wing-loading, and TOFL are negotiable differences in both aircraft variants. Previous data provides clear insights

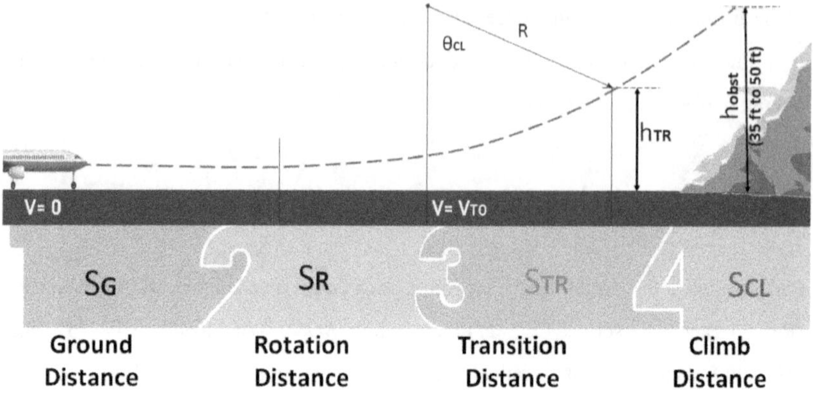

Figure 2.8: Gross takeoff distance

that wide-body aircraft is applicable for long-range with large passenger capacity. To propel such a large number of passengers requires sufficient thrust by propulsive systems. Thus, these systems indicate the number of engines, static thrust, thrust-to-weight ratio, fuel economy, and Thrust thrust-specific Fuel Consumption (TSFC). The majority of narrow-body aircraft have twin engines because of their lower capacity.

$$ASK = n_{PL} * R \tag{2.2.22}$$

$$S_{TO} = S_G + S_R + S_{TR} + S_{CL} \tag{2.2.23}$$

where, Ground distance, $S_G = \frac{V_{TO}^2}{a}$ and ground rolling acceleration, $a = \frac{g}{w}[T - D - \mu(W - L)]$

Rotation distance, $S_R = t_R V_{TO}$; for commercial flights, t_R ranges from 3 to 6 sec.

Transition distance, $S_{TR} = R sin\theta_{CL}$ where, transition height, $h_{TR} = R(1 - cos\theta_{CL})$ and climb angle = $\theta_{CL} = sin^{-1}\left(\frac{T}{W} - \frac{1}{\frac{L}{D}}\right)$

The rate of climb (RoC) stands as a crucial performance parameter, reflecting the interplay of various flight variables. In the depicted equation, an elevation in velocity and climb angle within the designated design results in an increased rate of climb as shown in Figure 2.9b. Conversely, an augmentation in thrust-to-weight ratio facilitates a reduction in drag-to-weight, thereby contributing to an enhanced rate of climb as shown in figure 2.9a. Climb distance = $S_{CL} = \frac{15.2-h_{TR}}{tan\theta_{CL}}$

$$RoC_{max} = \sqrt{\frac{\left(\frac{W}{S}\right)Z}{3\rho C_{D,0}}}\left(\frac{T}{W}\right)^{3/2}\left[1-\frac{Z}{6}-\frac{3cos^2\theta_{CL}}{2Z\left(\frac{T}{W}\right)^2\left(\frac{L}{D}\right)^2}\right] \tag{2.2.24}$$

where, $Z = 1+\sqrt{1+\frac{3}{\left(\frac{T}{W}\right)^2\left(\frac{L}{D}\right)^2}}$

Table 2.12: Propulsive operation of Wide-bodies and Narrow-bodies aircraft

Measures	Parameters	Wide-body aircraft	Narrow-body aircraft
Performance	(Vc)max (km/hr)	850 – 1100	777 – 895
	R (km)	6500 – 20372	2400 – 12205
	Endurance (hr)	6.9 – 19.7	2.9 – 14.6
	Service ceiling (ft)	31000 – 45100	28000 – 41000
	ROC (m/s)	5.08 – 28.96	5.08 – 15.28
	Wing loading (kg/m^2)	597 – 838	480 – 728
	LTO (m)	1700 – 3310	1770 – 3550
	ASK (seat.km)	1787500 –13136200	334712 –3219000
Propulsive system	No. of engine	2 – 4	2 – 4
	Engine Manufacturer	PW,GE, R-R	CFMI, PW, GE, R-R
	Total Static Thrust (kN)	510 – 1400	137 – 547
	Thrust –to-weight ratio	0. 23 – 0.31	0.22 – 0.34
	Fuel economy (km/litre)	0.05 – 0.13	0.07 – 0.78
	TSFC (kg/hr/N)	0.007 – 0.020	0.01 –0.03

$$S_{LO} = S_A + S_F + S_{FR} + S_{BR} \tag{2.2.25}$$

where, Approach distance, $S_A = \frac{15.3-h_F}{tan\theta_a}$; Flare distance, $S_F = Rsin\theta_{CL}$ where, transition height, $h_F = R(1-cos\theta_a)$ and approach angle = $\theta_a \leq 3$

Free roll distance, $S_{FR} = t_{FR}V_{LO}$; for commercial flights, t_R ranges from 3 to 6 sec.

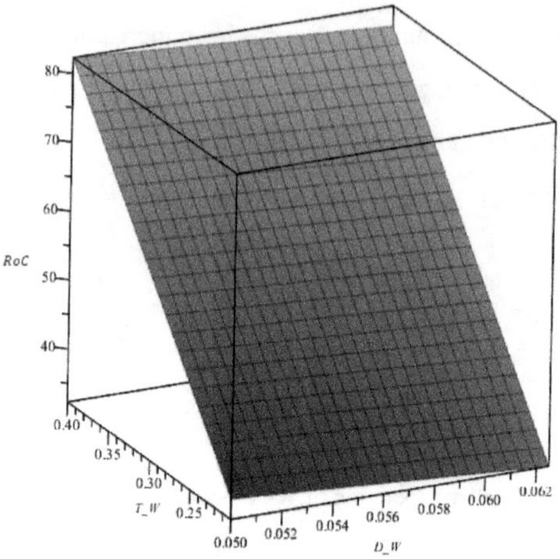

(a) Effect of RoC w.r.t Thrust and drag over Weight

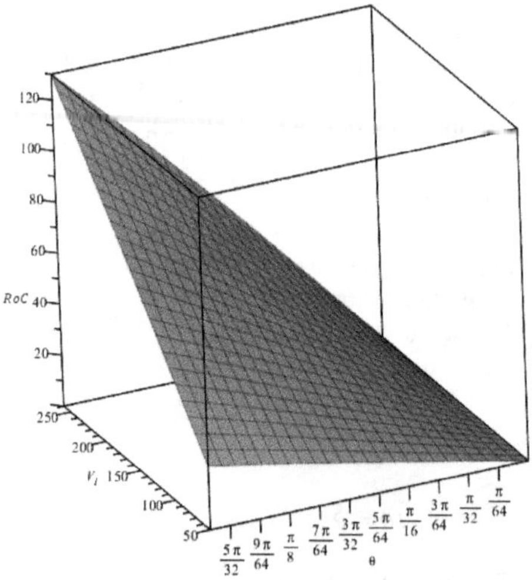

(b) Effect of RoC with Velocity and climb angle

Figure 2.9: Graphical representation of RoC with other performance parameters

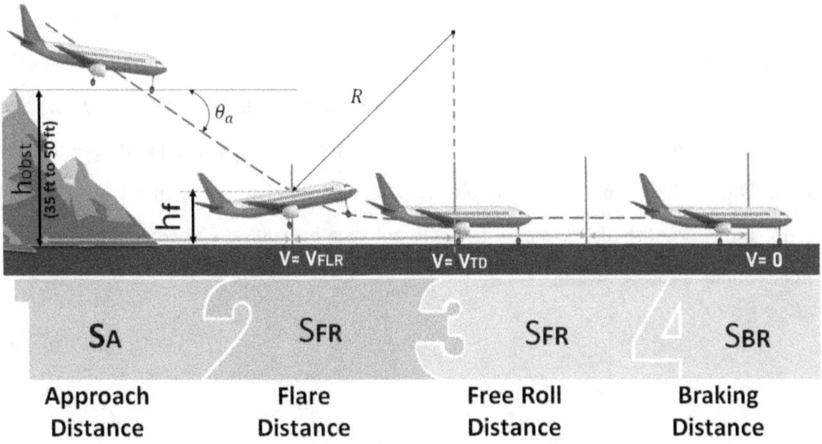

Figure 2.10: Total Landing distance

Braking distance = $S_{BR} = -\frac{V_{LO}^2}{r}$ and ground rolling retardation, $r = -\frac{g}{w}[T - D - \mu(W - L)]$

Thrust has a direct connection with drag produced in aircraft. If thrust is produced greater than drag initiates to propel the aircraft efficiently. Thrust requirement is strongly associated with aerodynamic features, structural integrity, and operating factors. Survey result declares a majority of the wide-body aircraft consists of four engines and newer models used only two turbofan engines. The market analysis says General Electric (GE) Aviation contributes 51% of turbofan engines for wide-body aircraft, and CFM International (CFMI) contributes 71% of engine shares for narrow-body aircraft. Remaining engine manufacturers like Rolls-Royce (R-R), Pratt & Whitney (P&W), and Engine Alliance contribute engine shares [87]. The thrust of each engine might vary in accord with the passenger capacity referred to as MTOW. According to correlation analysis, MTOW has a significant impact on fuel economy. IATA [80] reported that upgrading engines cut down fuel consumption by 10 percent [142]. Fuel efficiency improves by lower TSFC as shown in equation (2.2.26). The survey says wide-body aircraft provide lesser TSFC compared to narrow-body airplanes as shown in figure 5. The thrust-to-weight ratio signifies the operational factors based on the aircraft's mission profile. This indicator shows the performance in terms of improvement in ROC, aerodynamic efficiency, and drag-reduction [77]. Narrow-body and wide-body aircraft follow a standard range of thrust-to-weight ratio. Improving the fuel economy needs to enhance propulsive efficiency through BPR, noise, and emission as well [87] [145].

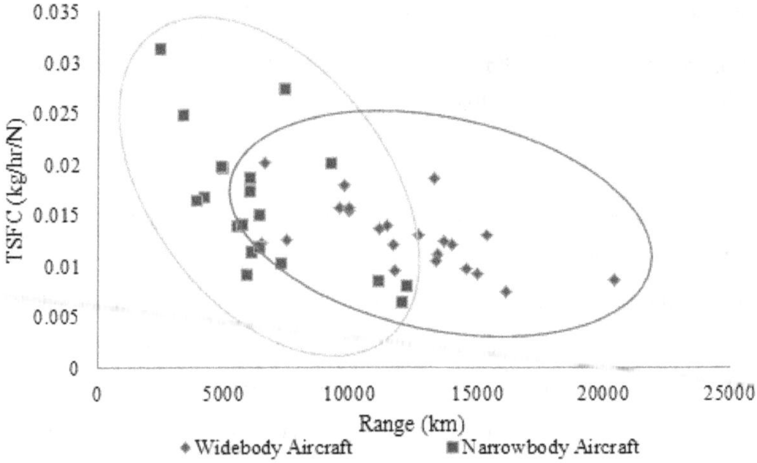

Figure 2.11: Variation of TSFC for different aircraft configuration

$$TSFC = \frac{\overline{W_{fuel}}}{T} \tag{2.2.26}$$

Where W_{fuel} is the Mass flow rate of fuel; T is the total thrust of the engine. The efficient design of aerodynamic surfaces in an aircraft is essential for achieving optimal performance, which includes reducing fuel consumption. Among these surfaces, the wing is of paramount importance as it primarily contributes to the force that lifts the aircraft and enables it to fly. However, the generation of lift by the wing isn't solely about getting the aircraft off the ground; it also has a significant impact on the overall balance and stability of the aircraft. This is where the horizontal tail plane (HTP) comes into play. Maintaining longitudinal trim is essential for stable flight. Longitudinal trim means ensuring the aircraft maintains a level flight path without continuous pitch adjustments. To achieve this, a counteracting force or moment is required to balance the moment generated by the wing. By adjusting the angle of the HTP's control surfaces (elevator or stabilator), the pilot or flight control system can control the aircraft's pitch and maintain stability. The HTP's ability to balance the moment generated by the wing is crucial for the aircraft's overall balance and stability. It ensures that the aircraft can maintain a desired attitude and altitude without continuous manual control inputs through tail volume coefficient (2.2.27), fin volume coefficient (2.2.28), Static margin (2.2.29), and damping ratio (2.2.30). This balance and stability are essential for the safety and efficiency of the aircraft.

$$V_{HT} = 0.46 * \left(\frac{L_f S_{HT}}{S_W C_{mac}} \right) \tag{2.2.27}$$

$$V_{VT} = 0.43 * \left(\frac{L_f S_{VT}}{S_W b_W} \right) \tag{2.2.28}$$

$$S.M = \overline{X_n p} - \overline{X_c g} \tag{2.2.29}$$

$$\zeta_{ph} = \frac{1}{2\sqrt{L/D}} \tag{2.2.30}$$

2.2.4 Socio-economical concerns

For consumers, manufacturers, and passengers, safety is a top priority. A transport aircraft is a critical aspect for passenger travelers who are concerned about safety and security. Unlike normal aircraft, the BWB features a unique fuselage structure that makes an emergency evacuation in 90 seconds difficult. The International Civil Aviation Organization (ICAO) is in charge of aviation safety and operating rules and regulations for aircraft manufacturers, maintenance processes, operating crew routines, and instruction. After each event, these regulations tightened. The Global Aviation Safety Plan (GASP) is to limit the incidence of risk by implementing security tactical planning [2]. Technical fault, spillage, explosion, and toxic gases are all common triggers for investigation. The failure of various aircraft controls might create a substantial threat to the public at any moment throughout the aircraft. More data on aviation security, mishaps, and incident investigations may be found at the Aviation safety network (https://aviation-safety.net/airlinesafety/enforcement/). In the last 11 years, incidents have been recorded, and it has been observed that the accident rate will fall by 0.9 in 2020. In a nutshell, improved aiding innovation such as glass cockpit can reduce the danger element. According to Wright[ ], engineering design has a direct impact on cost. As illustrated in the figure, costs in the aviation sector may be divided into two categories: direct operating costs and indirect operating costs. The cost of the aircraft depends on the mission in terms of fuel consumption. Short-haul missions have greater consumption of fuel than medium or long-haul missions [ ]. The study says that short-haul aircraft accessibility and levels of production are reduced compared with medium and long-range flights. It is also observed that a short-distance operation requires greater maintenance expenses, which results in lower profitability. Thus, maintenance cost or operating cost affects the design characteristics in five different ways are as follows: aircraft efficiency, engine number, size, speed, and aircraft age. According to Wesseler, the cost is determined by the aerodynamic profile. Improving fuel economy can be accomplished by optimizing overall power and minimizing the resistance of airflow accessories. The power plant and its force during the flight are the primary sources of fuel efficiency [ ]. Fuel is a significant expense in aviation which is subject to major industrial attempts to seek efficiencies [ ]. Fuel efficiency has a direct influence on the aerodynamic profile and the payload in terms of seat capacity, depending on the cost involved per passenger [107]. According to Hassan et al., [82] stated that fuel consumed (Wfuel) is a function of empty weight, MTOW, range (R), and LTO cycle as expressed in equation (2.2.31).

$$W_{fuel} = f(W_E, W_{TO}, R, LTOcycle) \tag{2.2.31}$$

In the year 2015, statistics said each one percent weight reduction corresponded to an increase in 0.75% fuel economy leading to increased coverage [17]. In practice, high-speed leads to improved profitability for certain seat capacities. If the planes are new then their life cycle and fuel efficiency will be improved [34]. According to Lee & Mo reported [107], the Airbus A380 appears to be more fuel-efficient than the Boeing B747-400. The gross weight of the airplane may grow as the number of passengers increases. The weight of an aircraft has a modest relationship with fuel economy. As a result, as a developer, MTOW is one of the essential characteristics that should be lowered by modifying the light structure and materials. However, it can increase effectiveness by extending the aircraft's distance and lowering its ecological footprint. The majority of the aircraft owners or buyers choose the aircraft based on three factors less fuel consumption, lowering the cost of the aircraft, and low emission and noise produced in it. Since then, the interest in business jets has been not only driven by innovation and development but also by market demand. Manufacturers gain competitiveness by assessing aviation market segments. Duong and Thao [64] emphasized market segmentation which helps to grow a firm's market by enhancing consumer satisfaction; enhancing profit or effectiveness, and creating relevant products or services based on the market segment. Interestingly, quality of service includes reliability, punctuality, scheduling, booking service, and boarding process such factors plays an important role [62] [63] [60] [185] [187]. At the same time, safety aspects and passenger comfort cannot be overlooked [152][133] [11]. In regard to avia-

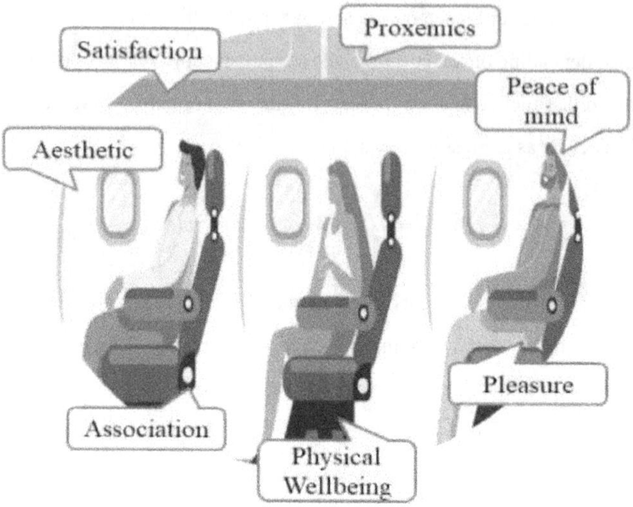

Figure 2.12: Elements of comfort factors [ ]

tion safety aspects organizational aspects (strategy and policy), human aspects (maintenance and operations), and operational aspects (Accident rates, risk management, air traffic management) refer to a multitude of factors, starting from security checkpoints as well as turbulence or airplane mechanical or technical issues [ ] [ ] [ ]. One of the essential factors in a business jet is passenger's comfort which follows odor, light, vibrations, acoustic, thermal impact, and anthropometry [ ] [ ] [ ]. Dhara et al., [ ] reported that passengers' comfort depends on their peace of mind, physical well-being, proxemics, joy, satisfaction, aesthetics, and association [ ] [ ] [ ] as shown in Figure 2.12 when choosing an airline company also affected by physical, and psychological, objects, and environmental aspects comprising cabin elements, social and situational. Bubb [ ] reported from the survey of 183 aircraft passengers that most ranked anthropometry as a key factor whereas other parameters ranked as noise, smell, climate, vibration, and light sequentially. Testimony of Robert Wilson from USITC Section 332 states that the major selection factors for a passenger aircraft are the aircraft's speed and range; the aircraft's cost; the architecture of the aircraft's passenger cabin; and the quality of customer [ ], or service [ ]. A real-time wide-body aircraft study is made to ensure that passenger comfort relates to the aisle width, seat width, and pitch width. Mathematically, FR value depends on the type of cabin configuration class as first-class, business class, economy class, and high-density class where a fuselage interior analysis can be obtained from the equation 2.2.32 and 2.2.33. A tradeoff study is necessary based on the passenger capacity that needs to be designed [ ] and is tabulated in Table 2.13.

$$D_f = (N_{abreast} * W_{seat}) + (N_{aisle} * W_{ailse}) + 2 * W_{thickness,fus} \tag{2.2.32}$$

$$L_f = L_{nose} + L_{cabin} + L_{rear} = L_{nose} + (N_{pitch} * W_{pitch}) + L_{rear} \tag{2.2.33}$$

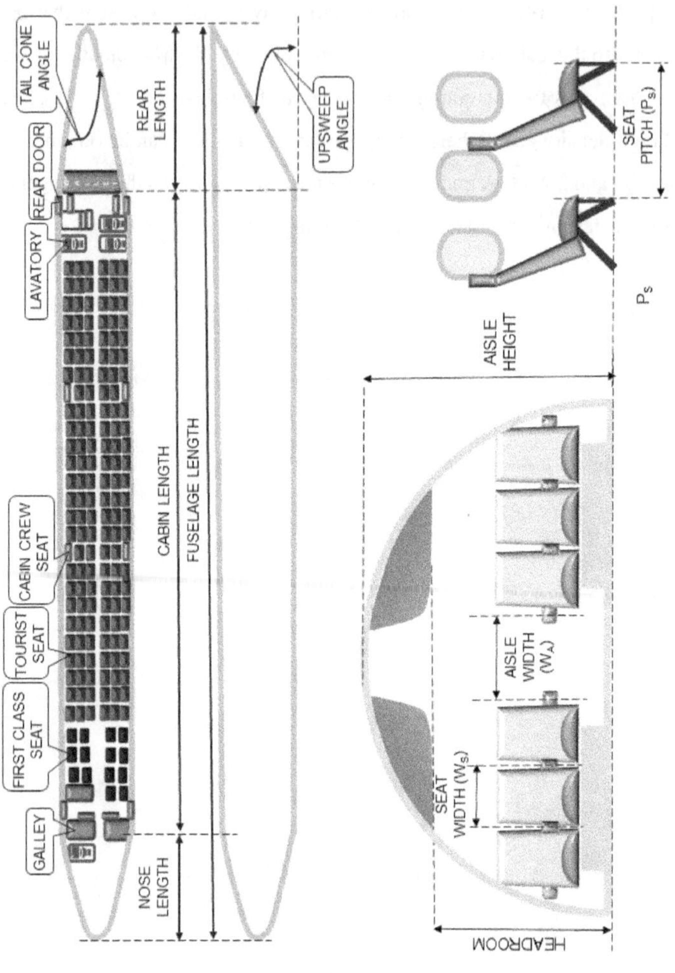

Figure 2.13: Overview of Fuselage interior

Table 2.13: Passenger arrangements and amenities in wide-body aircraft

Manufacturer	Configuration		Nabreast	Seating arrangements			In-flight amenities				
	Lf	FR		Wseat	Wpitch	Wailse	Audio	Visual	AC Power	Internet	food
	(m)			(cm)	(cm)	(cm)					
AIRBUS A300-600	53.300	9.450	10	43.942	57	38.1	yes	yes	yes	No	yes
AIRBUS A310-200	46.660	8.270	8	43.94	58	38.1	yes	yes	yes	No	yes
AIRBUS A330-900	63.660	12.100	10	45.16	60	43.18	yes	yes	yes	yes	yes
AIRBUS A330-300	62.470	11.076	10	43.94	62	41.91	yes	yes	yes	yes	yes
AIRBUS A340-300	62.470	11.076	10	43.30	68	38.1	yes	yes	yes	yes	yes
AIRBUS A340-600	69.570	12.335	8	43.99	56	38.1	yes	yes	yes	yes	yes
AIRBUS A380-800	72.720	10.180	9	43.99	55	43.18	yes	yes	yes	yes	yes
AIRBUS A350-900	66.800	11.910	8	38.86	58	40.64	yes	yes	yes	yes	yes
AIRBUS A350-900	73.790	12.110	10	43.94	57	43.18	yes	yes	yes	yes	yes
BOEING 747-100	68.630	10.558	10	45.97	55	43.18	yes	yes	yes	yes	yes
BOEING 747-200	68.630	10.558	10	43.942	56	39.624	yes	yes	yes	yes	yes
BOEING 747- 400	68.630	10.558	10	43.942	62	40.64	yes	yes	yes	yes	yes
BOEING 777-200	62.780	10.126	10	43.3	67	43.18	yes	yes	yes	yes	yes
BOEING 777-300	72.880	11.755	10	45.72	68	43.18	yes	yes	yes	yes	yes
BOEING 787-9X	68.280	11.830	9	43.942	58	40.64	yes	yes	yes	yes	yes
BOEING 777-9X	76.730	12.380	10	45.72	60	38.1	yes	yes	yes	yes	yes
DOUG. DC10-30	51.970	8.633	9	41.402	56	40.64	yes	yes	yes	yes	yes
McDON. MD - 11	58.650	9.743	11	43.942	67	40.64	yes	yes	yes	yes	yes
LOCKHD L1011-100	54.200	8.944	9	41.402	60	45.72	yes	yes	yes	yes	yes
ILYUSHIN II-86	56.100	9.227	10	41.402	68	42.93	yes	yes	yes	yes	yes
ILYUSHIN II-96-300	51.150	8.413	9	43.942	58	39.878	yes	yes	yes	yes	yes
ILYUSHIN II-96	60.500	9.951	9	43.942	56	39.878	yes	yes	yes	yes	yes

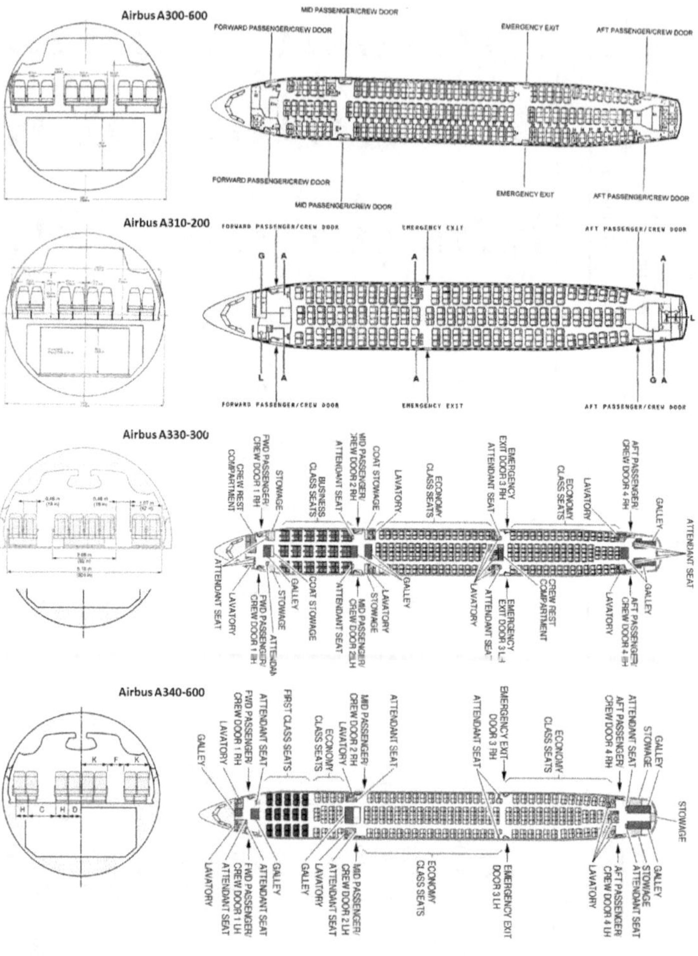

Figure 2.14: Widebody fuselage interior of A330-600, A310-200, A330-300, and A340-600 [ ]

In the given equation 2.2.32, with a fixed fuselage aisle configuration and aisle width, alterations in seat dimensions lead to variations in the number of seats, conse-

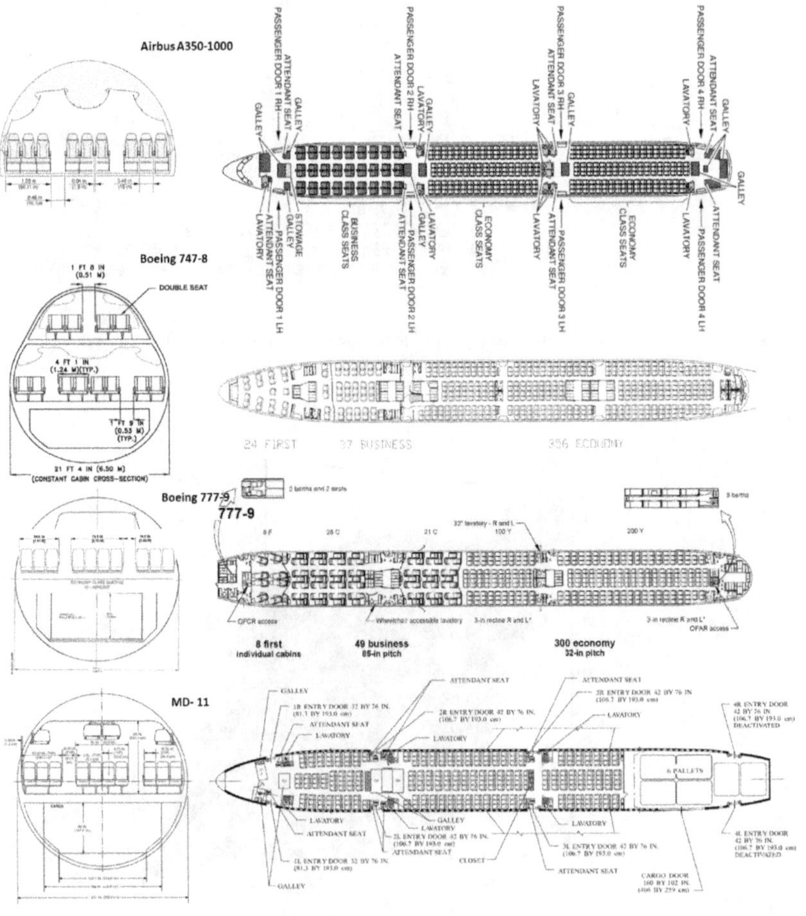

Figure 2.15: Different wide-body fuselage interiors [ ] [ ]

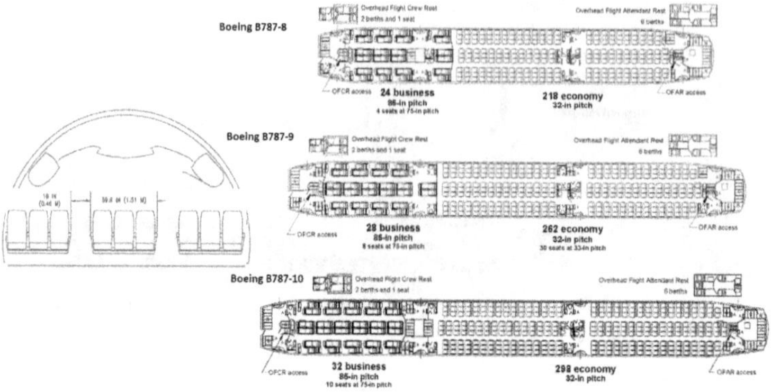

Figure 2.16: Boeing B787 series seating arrangements [ ]

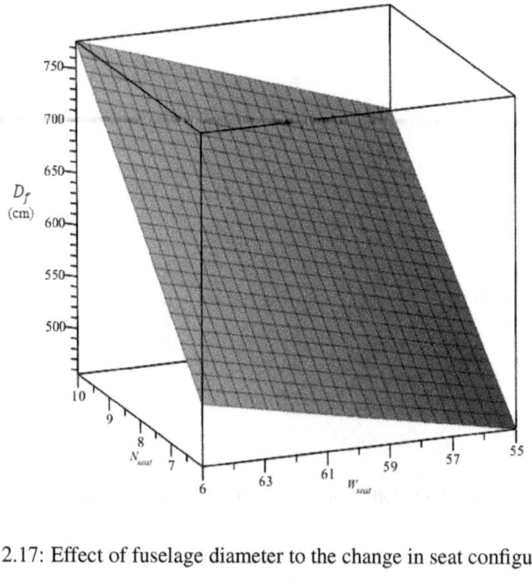

Figure 2.17: Effect of fuselage diameter to the change in seat configuration

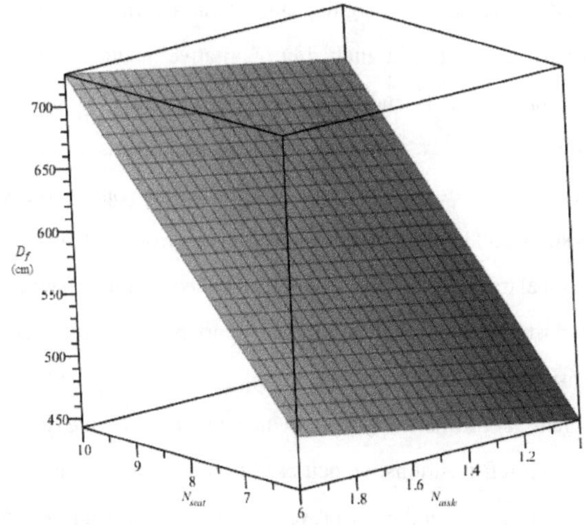

Figure 2.18: Effect of fuselage diameter with change in seat and aisle

quently influencing the fuselage diameter, as depicted in the accompanying figure 2.17. As the number of seats increases with a reduction in seat width, the fuselage diameter expands accordingly; conversely, a decrease in seat count with wider seats leads to a reduction in fuselage diameter. Contrary to the previous scenario, when maintaining fixed seat and aisle widths but adjusting the number of seats and aisles, a noticeable change in fuselage diameter is observed in Figure 2.18 . Specifically, a decrease in the number of seats and aisles results in a corresponding reduction in fuselage diameter [147].

2.3 Correlation Analysis

Building an aircraft is an intricate and multifaceted process involving a multitude of components, systems, and technologies. The term "complex system construction" underscores

the challenges and intricacies of designing and manufacturing aircraft. The statistical technique is a fundamental step in analyzing qualitative or quantitative data. This approach not only enhances safety and operational performance but also has a significant positive impact on the efficiency, profitability, and customer satisfaction levels of aviation companies. In the lack of established links that represent contemporary technical improvements, this study collects and organizes parameters from both historical and modern local and international transport aircraft to examine the correlations between these factors. In the aviation industry, safety is of paramount importance. Air traffic controllers play a critical role in ensuring the safety of flights [ ]. To make informed decisions and prevent serious problems, controllers rely heavily on data. Data provides them with real-time information about aircraft positions, velocities, altitudes, and weather conditions. Accurate and up-to-date data is essential for preventing runway overruns and collisions[ ]. Runway overruns occur when an aircraft cannot stop within the available runway length, and collisions may happen if aircraft come too close to each other. Data on aircraft speed, approach paths, and runway conditions help air traffic controllers coordinate take-offs and landings to maintain safe separations and reduce the risk of accidents. Airports need data to determine where aircraft should be positioned for take-off and landing [ ]. This includes data about the aircraft's type, size, weight, and destination. Based on this information, controllers allocate runways and taxiways, ensuring an efficient flow of aircraft movements. This data-driven approach is essential for efficient and safe aviation operations and ensures that the decisions made are reliable, based on real-time information, and aligned to prevent serious problems in the aviation industry [ ] [ ].
According to Profillidis and Botzoris [ ], the statistical tool is a key to defining transport modeling. This research focused on using the correlation coefficient technique. The main aim of correlation-based research is to identify the relationship between the variables. Such measuring values range from -1 to 0 to +1 illustrated in Table 2.14. By those values, identified how strongly the variable influences other parameters. The CORREL

function in MS Excel is employed in this article to determine the connection between the critical parameters. Critical parameters like aerodynamic configuration (Fuselage Length, wing Area, Wingspan, Aspect Ratio), structural integrity (MTOW, Payload mass), and performance and operational factors (Max. cruise speed, range, altitude, fuel economy, cost, thrust) correlation relationship studied among them.

In the aerospace industry, especially in aircraft design, numerous variables and parame-

Table 2.14: Correlation coefficient relationship study on aircraft parameters

	FR	S	b	AR	Λ	W_{TO}	$\frac{W_E}{W_{TO}}$	$\frac{W_{PL}}{W_{TO}}$	n_{PL}	W/S	T/W	LTO	Vc	R	ha	RoC	T	ASK	SFC
Lf	A+	A	A+	C+	C+	A	D	D	A	C+	D	A	A	A+	C+	C+	A+	A+	B
FR		C+	A	A	D	C+	C+	C+	B	D	D	C+	B	A	D	D	B	A	B
S			A	D	A	A+	B	C+	A	C+	D	A	C+	B	B	B	A	A	C+
b				A	D	A	A+	B	C+	A	C+	D	A	C+	B	B	B	A	A
AR					A	C+	D	C+	D	C+	D	C+	D	B	D	B	D	C+	A
Λ						B	C+	D	B	B	D	A	B	D	C+	A	B	C+	D
W_{TO}							D	C+	A+	B	C+	A	A	A	C+	B	A+	A	C+
$\frac{W_E}{W_{TO}}$								A+	D	D	A	C+	C+	C+	C+	D	D	C+	C+
$\frac{W_{PL}}{W_{TO}}$									C+	D	A	D	B	C+	C+	D	C+	C+	B
n_{PL}										C+	C+	B	A	B	D	B	A+	A	D
W/S											D	A	C+	C+	C+	D	B	C+	D
T/W												D	D	C+	D	D	D	C+	B
LTO													C+	B	C+	B	A	A	C+
Vc														B	D	C+	A	A	D
R															C+	D	A	A+	B
ha																D	B	C+	C+
RoC																	B	C+	D
T																		A+	C+
ASK																			B
R^2	0.993																		
Data	22																		

ters can affect the performance, efficiency, and safety of aircraft. These factors can range from the design of the wings and engines to the choice of materials and avionics systems. Regression analysis allows engineers and researchers to identify which of these factors significantly impact various data insights. Regression analysis is a statistical technique used to assess the relationship between a dependent variable (the outcome or result) and one or more independent variables (factors or parameters that can influence the outcome). In the context of aircraft design, the dependent variable might represent a critical performance metric such as fuel efficiency, load-carrying capacity, or structural integrity [70].

Regression analysis helps determine the significance of each independent variable. It quantifies how much each factor contributes to the variation in the dependent variable as mentioned in Table 2.14. This is crucial for prioritizing which parameters are the most influential in achieving a specific outcome are represented in the table. By using regression analysis to understand the compatibility and correlations of parameters, aerospace engineers can optimize aircraft design. It aids in understanding which factors are critical in achieving specific outcomes, and its results can contribute to informed decision-making, optimization, and addressing the challenges in aerospace design and engineering.

2.4 Summary

According to Boeing's statement, wide-body aircraft configuration is the future to generate market competitiveness for the airlines. Designers depend extensively on historical design expertise and an extensive collection of statistical data throughout the conceptual design stage of aircraft development, frequently relying on established empirical methods for measuring overall and performance metrics [[illegible]]. Further investigation of the design parameters was carried out from previous literature as mentioned in Table 2.2.

This chapter provides the answers to research questions that are connected to each of the objectives the most important elements influencing the design and performance of wide-body aircraft fuselage. The next chapter provides a clear insight into the effective Multi-Criteria Decision-Making (MCDM) Techniques to select an optimized wide-body aircraft systematically.

Chapter 3

Selection of wide-body aircraft through Multi-Criteria Decision-Making(MCDM)Techniques

3.1 MCDM: Background

As an aircraft designer, future air transportation needs to be sustainable by improving efficiency, economic competitiveness, lowering emissions, and enhancing reliability [48][ ]. According to IATA [86], progressive technologies such as green aviation technology may maximize better fuel economy per passenger by 70 percent. Aircraft design keeps track of parameters related to performance, such as speed, range, altitude, take-off distance, wing loading, and thrust loading [72][154]. Besides, it evaluates aircraft reliability and profitability and estimates accident rates, safety features, costs, revenue generation, and fuel economy. However, selecting the optimized MoM aircraft from the designer's or buyer's point of view is quite tedious [15]. This problem can be solved using multiple criteria decision-making (MCDM) to aid design professionals, in selecting appropriate or suitable aircraft reaching more suitable selections for certain operations [124] [94] [157]

[ ] [ ].

Any new aircraft design and development evolved with multiple stages. Therefore, to achieve a good design, each step demands standards governing and suitable decisions as per requirement [ ] [ ] [ ] [ ]. Strategic planning in these phases of aircraft design frequently involves numerous criteria, which is critical to employ multiple criteria decision-making (MCDM) to aid design professionals in reaching more suitable selections. There are popular MCDM approaches such as Analytic Network Process (ANP) [ ], Analytic Hierarchy Process (AHP) [ ][ ] [ ] [ ]; [ ], Entropy weight method (EWM) [ ], Technique for Order of Preference by Similarity to Ideal Solution (TOPSIS) [ ] [ ] [ ] [ ] [ ], VIseKriterijumska Optimizacija I Kompromisno Resenje (VIKOR) [ ], Simple Additive Weighting (SAW) [ ], Evan Swaps Method (ESM) [ ] Multi-Objective Optimization Method by Ratio Analysis (MOORA) [ ], Novel Approach to Imprecise Assessment and Decision Environments (NAIADE) [ ], ELimination Et Choice Translating Reality (ELECTRE) [ ], Reference Analysis for Reference Ideal solution (PARIS) [ ], Fuzzy Reference Ideal Method (FRIM) [ ] [ ], Weighted Aggregates Sum Product Assessment (WASPAS) [ ] [ ] and some more methods.

Table 3.1 represents a contribution of the MCDM method in the aerospace industry for the selection and evaluation of transport, fighter, military, Unmanned Aerial Vehicles (UAV), and Regional aircraft. These techniques help to select and evaluate the appropriate aircraft based on the requirement. MCDM method helps to characterize the sorts of decision-making challenges that come throughout the various design phases, technical, financial, and safety strategies for solving issues incorporated in commercial or defense [ ] [ ] [ ]. Some literature also worked on MCDM techniques used in airline quality services and passenger comforts [ ] [ ] [ ] [ ] [ ]. Evaluating technical parameters and conflicting aspects among the alternatives can be solved using a suitable MCDM approach. A benefit of the MCDM approach aims to predict both measurable and subjective factors using single and hybrid techniques. There are ample MCDM techniques used for the engineering selection process, technical aspects, and operational research [120]. Table 3.1 clearly states an implementation of technical aspects and variants sub-parameters are considered based on the objective finding. The MCDM method helps to characterize the sorts of decision-making challenges that come throughout the various design phases, and technical, financial, and safety strategies for solving issues incorporated in commercial or defense [55] [79] [134]. Some literature also worked on MCDM techniques used in airline quality services and passenger comforts [56] [196] [60]. Evaluating technical parameters and conflicting aspects among the alternatives can be solved using a suitable MCDM approach [133]. A benefit of the MCDM approach aims to predict both measurable and subjective factors using single and hybrid techniques. At this point, a novel conceptual design needs to develop the flight mission and evaluate the result. It shows a wide opportunity and challenge of execution of MCDM to select or evaluate the appropriate aircraft based on the objective function.

3.2 Proposed Approach

According to Boeing's statement, wide-body aircraft configuration is the future to generate market competitiveness for the airlines. In this section, data were collected based on 22 real-time wide-body transport aircraft, which includes a detailed study of 'common-to-all types' such as retrofit configuration, propulsive systems, structural parameters, and operational performances. At this point, a novel conceptual design needs to develop the flight mission and evaluate the result. In consequence, the proposed method is segmented into four steps: Firstly, choose the BASELINE aircraft based on fuel economy. The role of decision-making strategy plays a vital role in choosing the right aircraft as the base-

Table 3.1: Application of MCDM method in the aerospace industry

Authors Reference	MCDM Method used	Criteria			Findings
		Technical Aspects	Economic Aspects	Safety Aspects	
[171]	ELECTRE, TOPSIS, SAW	Speed; MTOW SFC; Npax	ASK	-	Evaluating the apt MCDM method of selecting aircraft
[154]	AHP	-	Price; Cost;	Security; Reliability; service quality	Evaluating the systems performance
[75]	Fuzzy Stochastic	Range; Speed; LTO cycle; Dimensions	DOC	Avionics availability; Comfort; Safety;	Assessing the aircraft based on performance, economic and safety concerns
[160]	Fuzzy AHP TOPSIS	Hopper capacity; LTO cycle; Fuel capacity; Engine Power; Aspect Ratio; Rateof climb; Dihedral; Wing Span; SFC	-	-	Selection of aircraft according to the considered criteria.
[61]	AHP	Seat capacity; MTOW; Baggage capacity;	Price; Fuel per seat; CASK	-	Selecting the. best aircraft based on the decision maker
[62]	AHP ESM	Seat capacity; MTOW; Baggage capacity;	Price; Fuel per seat; CASK	-	Comparative study of different MCDM technique on aircraft

[153]	FANP FAHP,	-	Price; Cost;	Security; Reliability; service quality	Comparison of different techniques and obtain the suitable result
[187]	APH TOPSIS	capacity; Maneuvering; performance	-	-	Selecting the appropriate aircraft For defense
[65]	Fuzzy AHP	Seat capacity; MTOW; Range;	Cost; CASK	Comfort;	Estimating the operational performance of the aircraft
[22]	FSAW, FTOPSIS, FVIKOR, ARAS-F, COPRAS-F, Fuzzy MULTIMOORA, Fuzzy DEA	-	Labor cost; Financial performance	Service performance	Evaluating the safety system and ways to improve the financial performance
[196]	APH TOPSIS	-	-	Registered; Accidents; Fatality rate;	Selecting the aircraft for flying training school
[157]	FRIM	Ceiling Endurance; Thrust; MTOW; Speed; LTO cycle; Range; Tactical capacity;	Cost	Ergonomic; Infrastructure	Selection of aircraft as per criteria

[[illegible]]	APH TOPSIS, VIKOR	Wingspan; Payload capacity, endurance; Speed; Altitude; range	-	-	Selecting and finding sensitivity analysis with different MCDM technique
[[illegible]]	Fuzzy APH TOPSIS	Seat capacity; MTOW; SFC; LTO cycle; Range; Speed	Price; Fuel per seat; DOC	-	Selecting best suitable commercial aircraft using MCDM
[[illegible]]	PARIS	MTOW; Speed Range; Service ceiling; Rate of climb; Seat capacity	-	-	Select the best aircraft among the alternatives using MCDM
[[illegible]]	Entropy WASPAS	Seat capacity; MTOW; SFC; LTO cycle; Range; Speed	DOC	-	Selection and evaluation process using MCDM
[[illegible]]	CRITIC- TOPSIS	MTOW; Speed; Range; Aesthetic;	-	-	Selection of aircraft and sensitivity analysis
This chapter	CRITIC- WASPAS, TOPSIS, SAW, WASPAS, TOPSIS, SAW	Fineness ratio; Aspect ratio; MTOW; Empty weight fraction; Seat capacity; LTO cycle; Speed; Range; Altitude; SFC	Fuel cost per seat;	No of accidents	Comparative study, Selection and sensitivity analysis

line aircraft. A literature review of several MCDM approaches in the aviation sector was conducted. The CRITIC-WASPAS MCDM approach was used in this study to select the optimum baseline aircraft of choice. Secondly, once the selection of baseline aircraft is accepted, according to the mission profile of the flight will be considered as a base flight mission. The baseline aircraft will determine to validate the result with a novel conceptual one in terms of a retrofit design structural contributions or operational performances. Thirdly, the indulging parametric changes and incorporating advanced technology to find the MTOW of the novel aircraft. The current manuscript focuses on MCDM techniques integrated with the interior point design optimization approach. In the proposed optimization technique, interior point methods have been adopted. Its solution can be either a global optimum or no feasible solution will exist. Finally, the results found in an optimized solution are reported and compared with the baseline discussed in this study. The robustness of the novel design was validated by comparing the operational performance of the baseline aircraft.

3.2.1 Hybrid Proposed Approach

This sub-section employing the critical parameters ought to find the baseline aircraft using the Multi-criteria decision-making (MCDM) approach. Dhara et al. [ ] insight into the design parameters affects the performance in terms of fuel economy or weight reduction. Table 2 clearly states an implementation of technical aspects and variants sub-parameters are considered based on the objective finding. It shows a wide opportunity and challenge of execution of MCDM to select or evaluate the appropriate aircraft based on the objective function. Based on the literature survey, the 'Criteria Importance Through Intercriteria Correlation' (CRITIC) approaches are not yet used in the aerospace sector to obtain the objective weights of considered criteria. 'Weighted Aggregates Sum-Product

Assessment' (WASPAS) Method proved to have a highly significant rate of accuracy [28]. This paper proposed a hybrid MCDM called the CRITIC-WASPAS approach applied to find the best and worst midsize aircraft as shown in Figure 3.1.

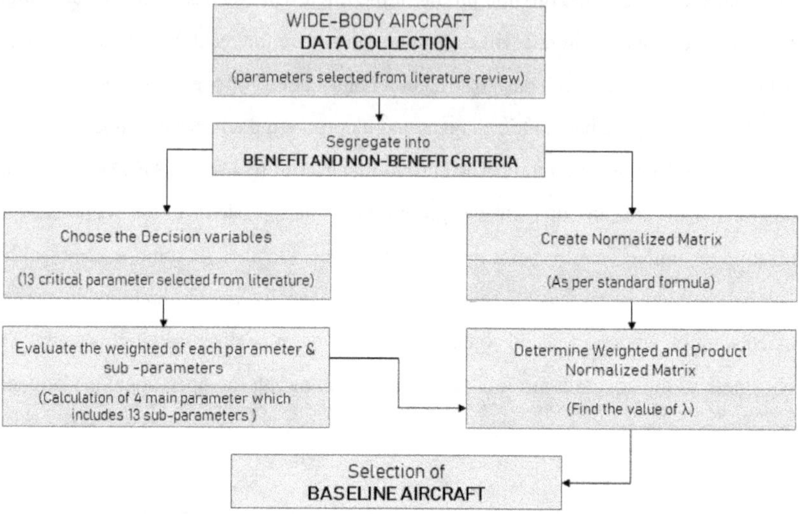

Figure 3.1: Process of choosing baseline aircraft

CRITIC approach

The main motive of the CRITIC approach is to find out the object weights of the criteria considered. The fundamental principle of this method is to measure conflicts through statistical data [38]. This approach provides the weighted priority based on the measures of conflicts through standard deviation for each criterion. A formulation of objective weights begins with four simple steps. Firstly, starts with creating a decision matrix, framing a normalized matrix based on beneficial and non-beneficial criteria, developing a correlation matrix, and applying an objective index to obtain weights of criteria.

Step 1: Create a Decision Matrix

The construction of a decision criteria matrix for wide-body transport aircraft is a combination of alternatives and criteria used in aircraft design. There are 22 different alternatives with 20 different criteria. The generalized criteria-decision matrix '$D_{m\times n}$' expressed in equation (3.2.1) is

$$D_{m\times n} = \begin{bmatrix} D_{11} & D_{12} & ... & D_{1n} \\ D_{21} & D_{22} & ... & D_{2n} \\ ... & ... & ... & ... \\ D_{m1} & D_{m2} & ... & D_{mn} \end{bmatrix} \tag{3.2.1}$$

Where, 'm' number of alternatives, and 'n' number of criteria.

Table 3.2: Decision criteria of wide-body transport aircraft

C1 Lf (m)	C2 FR	C3 S (m²)	C4 b (m)	C5 AR	C6 Λ (°)	C7 MTOW (kg)	C8 $\frac{W_E}{W_{TO}}$	C9 $\frac{W_{PL}}{W_{TO}}$	C10 nPL	C11 W/S (kg/m²)	C12 T/W	C13 LTO (km)	C14 Vc (km/h)	C15 R (km)	C16 ha (ft)	C17 RoC m/s	C18 T (kN)	C19 ASK	C20 SFC (hr^{-1})
53.3	9.5	260.0	44.8	7.7	28.0	170500	0.52	0.24	375	655.8	0.31	2280	897	7500	31000	16.3	514	2812500	0.013
46.7	8.3	219.0	43.9	8.8	28.0	144000	0.55	0.23	275	655.8	0.27	2280	850	6500	41100	16.3	515	1787500	0.012
63.7	12.1	191.0	64.0	8.5	28.0	227703	0.55	0.23	460	657.5	0.22	2280	1005	13334	33000	16.3	648	6133640	0.019
62.5	11.1	363.1	58.0	9.3	29.7	217000	0.54	0.22	440	597.6	0.28	2770	870	11760	33000	10.2	600	5174400	0.010
62.5	11.1	363.1	58.0	9.3	29.7	271000	0.48	0.18	440	746.4	0.23	3000	913	13700	33000	29.0	604	6028000	0.012
69.6	12.3	437.3	61.2	8.6	31.1	365000	0.48	0.17	475	834.7	0.28	3140	915	14600	41450	29.0	996	6935000	0.010
72.7	10.2	845.0	79.8	7.5	33.5	575000	0.48	0.15	853	680.5	0.25	3000	1099	15400	43000	7.6	1392	13136200	0.013
66.8	11.9	442.0	64.8	9.5	31.9	280000	0.41	0.19	440	633.5	0.27	2600	945	15000	43100	10.2	749	6600000	0.009
73.8	12.1	464.3	64.8	9.0	31.9	286670	0.49	0.22	480	633.5	0.27	2600	945	16100	41450	10.2	863	7728000	0.008
68.6	10.6	511.0	59.6	7.0	37.5	340195	0.50	0.20	516	665.7	0.25	3246	939	9800	35000	5.1	835	5056800	0.018
68.6	10.6	511.0	59.6	7.0	37.5	377840	0.47	0.17	516	739.4	0.25	3200	969	12700	45100	5.1	944	6553200	0.013
68.6	10.6	525.0	62.3	7.4	37.5	396830	0.46	0.15	660	755.9	0.26	3310	933	13450	35000	7.6	1010	8877000	0.011
62.8	10.1	427.8	60.9	8.7	31.6	286897	0.48	0.20	440	670.6	0.27	2438	950	14000	43100	15.2	747	6160000	0.012
72.9	11.8	427.8	60.9	8.7	31.6	299370	0.52	0.23	550	699.8	0.29	3230	950	11165	43100	15.2	846	6140750	0.014
68.3	11.8	377.0	60.1	9.6	32.2	254011	0.53	0.23	440	673.7	0.27	2800	954	11750	41100	15.2	680	5170000	0.012
76.7	12.4	516.7	71.8	10.0	31.6	352441	0.51	0.21	426	682.1	0.27	3000	1035	20372	43100	15.2	934	8678472	0.009
52.0	8.6	367.7	50.4	6.9	35.0	263636	0.46	0.13	380	717.0	0.27	3000	982	9630	42000	10.2	708	3659400	0.016
58.7	9.7	338.9	51.8	7.9	35.0	283720	0.47	0.20	405	837.2	0.30	3139	961	13408	43000	10.2	822	5430240	0.010
54.2	8.9	321.5	47.4	7.0	35.0	211374	0.53	0.16	400	657.5	0.27	2350	957	6667	33000	14.2	561	2666800	0.020
56.1	9.2	320.0	48.1	7.2	35.0	208000	0.40	0.20	350	650.0	0.25	2700	869	10000	43000	15.0	510	3500000	0.015
51.2	8.4	391.6	55.6	7.9	30.0	216000	0.54	0.19	300	650.0	0.25	2400	850	11500	43000	15.0	628	3450000	0.014
60.5	10.0	391.6	55.6	7.9	30.0	270000	0.49	0.21	375	650.0	0.25	2600	870	10000	43000	15.0	658	3750000	0.016

Applying equation (3.2.1), Table 3.2 represents the criteria decision matrix D for a midsize aircraft. Next is to calculate the linear normalized decision matrix by screening the benefit as the maximum value and non-benefit as a minimum value from the decision matrix Dij using equations (3.2.2a) and (3.2.2b) expressed as

$$\overline{N_{ij}} = \frac{D_{ij} - D_{ij,worst}}{D_{ij,best} - D_{ij,worst}} \tag{3.2.2a}$$

$$\overline{N_{ij}} = \frac{D_{ij,best} - D_{ij}}{D_{ij,best} - D_{ij,worst}} \tag{3.2.2b}$$

Where i = 1, 2 ,. . . ., m and j = 1, 2,.., n. Setting the Best and worst parameters, Table 3.3 represents the normalized decision matrix Nij and the standard deviation of each jth criterion from C1 to C13. Mathematically, the standard deviation (σ) can be derived through MS Excel STDEVPA of each criterion. Followed by evaluating the correlation coefficient relation matrix rij of each criterion by applying MS Excel operator CORREL. To predict the conflicting index and weights of each decision criterion, a sum of all coefficient relation matrix rij are tabulated in Table 3.3.

Step 2: Measures of conflict index (Ij) and weights (Wj). This step is to find out the value of conflict index Ij and objective weights Wj to obtain the weightage of each criterion using equations (3.2.3) and (3.2.4) and represented in Table 3.3. It is been noticed that aerodynamic consideration causes 19.41 percent contribution, structural consideration attributes 20.21 percent, Performance impacts about 28.55 percent, market profitability contributes 21.51 percent, and reliability provides about 10.32 percent.

$$I_j = \sigma \sum_{j=1}^{n} \left(1 - r_{ij}\right) \tag{3.2.3}$$

$$W_j = \frac{I_j}{\sum_{j=1}^{n} I_j} \tag{3.2.4}$$

Table 3.3: Measures of conflicts

Notations		Standard deviation (σ)	Conflict index (Ij)	Objective weights (Wj)	Rank
C1	Lf (m)	0.56	9.05	0.05	8
C2	FR	0.46	10.74	0.06	3
C3	S (m²)	0.52	9.07	0.05	7
C4	b (m)	0.57	9.93	0.06	6
C5	AR	0.38	6.67	0.04	17
C6	Λ (°)	0.44	8.48	0.05	11
C7	MTOW (kg)	0.58	12.49	0.07	1
C8	WE/WTO	0.41	8.26	0.05	12
C9	WPL/WTO	0.41	8.08	0.05	13
C10	nPL	0.51	11.11	0.06	2
C11	W/S(kg/m²)	0.36	6.69	0.04	16
C12	T/W	0.33	6.47	0.04	18
C13	LTO (km)	0.49	10.64	0.06	4
C14	Vc (km/h)	0.47	8.56	0.05	10
C15	R (km)	0.48	7.59	0.04	15
C16	ha (ft)	0.30	5.13	0.03	20
C17	RoC m/s	0.41	7.80	0.05	14
C18	T (kN)	0.58	9.96	0.06	5
C19	ASK	0.56	8.77	0.05	9
C20	SFC (hr^{-1})	0.383895	6.39845	0.03722	19

WASPAS Method

The implication of the WASPAS method was developed by Zavadskas et al. [[illegible]], the motto to obtain higher rank accuracy. This approach is a combination of two basic techniques called WSM and WPM. This approach portrays a benefit in rank accuracy, rationality, and practicality preceded by basic four steps are as follows:

Step 1: Obtain the Linear Normalized Decision Matrix.

The first step of the WASPAS method is to generate a linear normalized matrix $\overline{C_{ij}}$ based on benefit and non-benefit criteria on the jth column as shown in equations (3.2.5a) and (3.2.5b), respectively.

For beneficial criteria,

$$\overline{C_{ij}} = \frac{C_{ij}}{max_i C_{ij}} \tag{3.2.5a}$$

For non-beneficial criteria,

$$\overline{C_{ij}} = \frac{C_{ij}}{max_i C_{ij}} \tag{3.2.5b}$$

Where $\overline{C_{ij}}$ is the normalized value of D_{ij}.

Step 2: Computing the Normalized Weighted Matrix

. In this step, need to evaluate the weighted sum and product matrix implementing WSM and WPM. This value can be obtained from the mathematical expression in equations (3.2.6a) and (3.2.6b) respectively.

$$WSM = \psi_i^{(1)} = \sum_{j=1}^{n} \left(\overline{C}_{ij}\right) * w_j \tag{3.2.6a}$$

$$WPM = \psi_i^{(2)} = \prod_{j=1}^{n} \left(\overline{C}_{ij}\right)^{w_j} \tag{3.2.6b}$$

Where Wj refers to the objective weights obtained from the CRITIC method for each criterion. The weightage of C1 to C20 criteria is considered in Table 3.3. Implementation of the equations (3.2.6a) and (3.2.6b), are computed and tabulated in Table 3.4. According to Zavadskas et al., [190] introduced a generalized equation expressed as equation (3.2.7). This equation (3.2.7) is the combination of WSM and WPM used in equations (3.2.6a) and (3.2.6b)) with a WASPAS coefficient λ. The value of λ ranges from 0 to 1 (0, 0.1, 0. 2, ..., 1). The table 3.4 illustrates by compiling equation (3.2.7) varying the value of λ from 0 to 1 at a unit step of 0.1.

$$\psi_i = \lambda \psi_i^{(1)} + (1 - \lambda) \psi_i^{(2)} \tag{3.2.7}$$

This WASPAS technique is a unique combination of WSM and WPM methods. If the value of λ is 0, then ψ_i becomes $\psi_i^{(2)}$. Similarly, if the value of λ is 1, then ψ_i becomes $\psi_i^{(1)}$ i.e., equivalent to WSM.

Step 3: Evaluate the Rank of an aircraft

The values obtained for different values of λ from 0 to 1. The value of λ closest to 1 refers to the highest rank among the alternatives illustrated in Figure This phenomenon yields the rank for different values of λ. Considering the average value of all values will determine the best and worst selection of midsize aircraft By employing the WASPAS MCDM technique, the rank of midsize aircraft has been selected from Table 3.4 as follows:

λ = 0.5: A2 > A1 > A10 > A19 > A20 > A17 > A22 > A4> A11 > A14 > A15 > A12 > A21 > A5 > A18 > A7 > A13 > A8 > A9 > A6 > A16 > A3,
i.e., Airbus A310-200 > ... > Airbus A330-900.

3.3 Results and Discussions

This section is to find out the practicality and stochastic nature of the proposed CRITIC-WASPAS Approach in midsize aircraft. Addressing the CRITIC method, the midsize

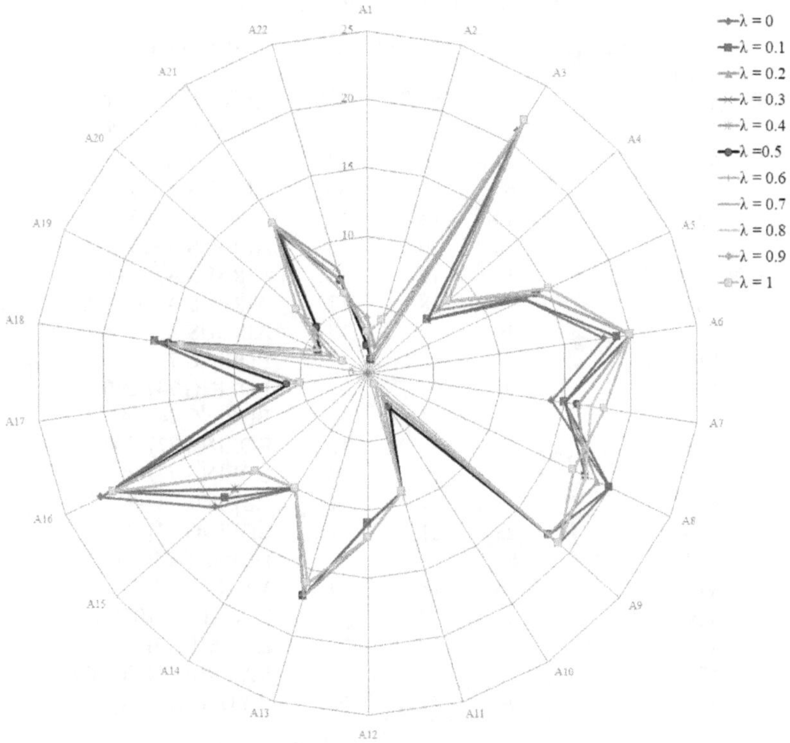

Figure 3.2: Proposed MCDM rank analysis

aircraft's case studies are made to interpret the design parameters and configuration to determine the objective weights of the criteria. In this work, all the twenty different objective criteria C1 to C20 are obtained from the CRITIC approach. As per the CRITIC method, fuel consumption (C20), criteria is the key design consideration followed by passenger capacity (C10), MTOW (C7), fineness ratio (C2), Thrust-to-weight ratio (C12), and other parameters. The objective weight criteria signify the importance of designing an optimization of the wide-body fuselage that relates vitally to aircraft design. This paper

Table 3.4: Computed the rank of WASPAS technique

Alternative decision variables	$\lambda = 0$	$\lambda = 0.5$	$\lambda = 1$	Model Aircraft
A1	2	2	3	AIRBUS A300-600
A2	1	1	2	AIRBUS A310-200
A3	21	22	22	AIRBUS A330-900
A4	6	8	8	AIRBUS A330-300
A5	13	14	15	AIRBUS A340-300
A6	18	20	20	AIRBUS A340-600
A7	14	16	18	AIRBUS A380-800
A8	20	18	17	AIRBUS A350-900
A9	19	19	19	AIRBUS A350-900
A10	3	3	1	BOEING 747-100
A11	9	9	9	BOEING 747-200
A12	11	12	12	BOEING 747- 400
A13	17	17	16	BOEING 777-200
A14	10	10	10	BOEING 777-300
A15	15	11	11	BOEING 787-9X
A16	22	21	21	BOEING 777-9X
A17	8	6	5	DOUG. DC10-30
A18	16	15	14	McDON. MD - 11
A19	4	4	4	LOCKHD L1011-100
A20	5	5	7	ILYUSHIN II-86
A21	12	13	13	ILYUSHIN II-96-300
A22	7	7	6	ILYUSHIN II-96

discerned the factors that influence aircraft design. Furthermore, the motive of this paper is to find out the rank of the considered alternative by applying the WASPAS approach in equation (3.2.7). An execution of equation (3.2.7) by varying the factor λ by 0.1 unit step yields Table 3.4. It reveals both the best and worst alternatives dominantly as A2 (Airbus A310-200) and A3 (Airbus A330-900 N) respectively. As per the WASPAS MCDM approach, the Airbus A310-200 (A2) aircraft alternative has obtained the first rank among the midsize aircraft in all three cases where the λ value is 0, 0.5, and 1 and the schematic dimensional model [ ] [ ] is shown in Figure 3.3. A strong indication of the worst aircraft is identified as the Airbus A330-900 N (A3) aircraft as per the considered objective

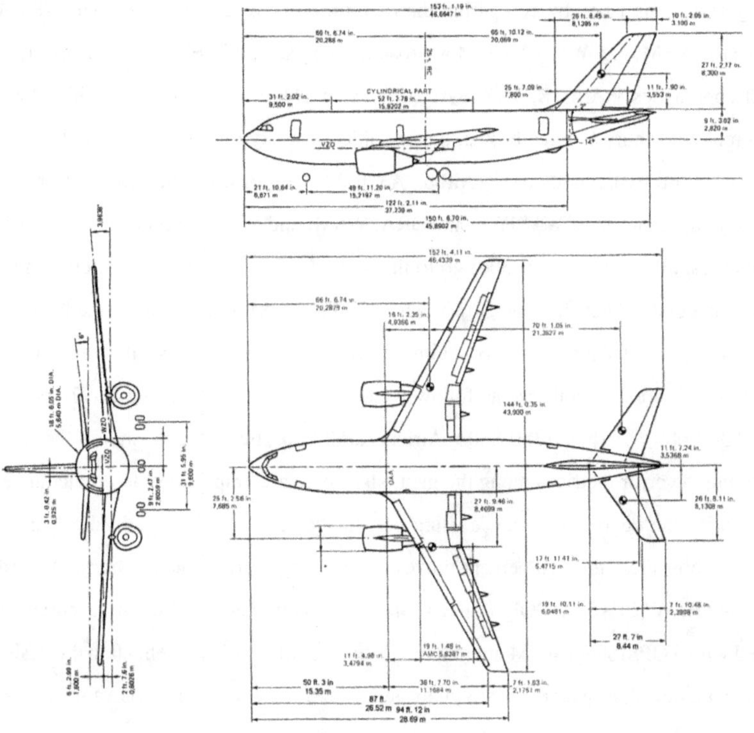

Figure 3.3: Dimensional model of best Alternative Airbus A310-200

criteria.

In addition, compare the rank using the WASPAS with other rank priority MCDM techniques to study the rate of accuracy of this approach. There are two basic ways to check the rank accuracy either by varying or fixing the objective criteria. This investigation is carried out only by using different MCDM techniques without affecting the object weight derived from the CRITIC method. To check the stochastic nature of the alternative analysis by a comparative study with other rank-based MCDM techniques. A comparative study is accomplished by comparing the rank order using the WASPAS with MOORA, TOPSIS, WSM, and WPM MCDM approaches represented in Figure 3.4. Thus, five distinct approaches WASPAS, MOORA, TOPSIS, WSM, and WPM are identified through different color bars as blue, brown, green, violet, and sky blue, illustrated in Figure 3.4. The aforesaid explains that A2 (Airbus A310-200) attained the first rank order for the majority approaches as WASPAS, MOORA, WSM, and WPM Approaches. In considerable cases, all (Airbus A330-300) go to the second rank order of better aircraft in terms of parameters and considerations. Similarly, in the majority of the cases, the A1 (Airbus A300-600) is considered the worst aircraft among the alternatives. It is observed that the top five worst aircraft are as follows: Boeing B747-100 > Boeing B777-9X > Airbus A340-600 > Airbus A380-800 > Airbus A350-900.This research extends the work to compute the error metrics among the methods. There are ample ways to predict the error to identify the degree of variance using percentage errors [ ]. The result of the error metrics shows less than 10 percent proves to be strongly error-free accuracy. According to sMAPE, WASPAS, MOORA, and WSM methods proved to be highly accurate compared with TOPSIS and WPM approaches. These results can be beneficial to the designer, engineers, decision-makers, or any managerial point of view to provide valuable insight to implement new design technology to improve flight performance.

3.4 Summary of key findings

This chapter presented a unique MCDM methodology to select the rank of midsize aircraft. The study of existing literature discovered numerous challenges regarding decision support for the technical design, financial, and safety strategies incorporated in aerospace industries. This research intended to obtain the most appropriate midsize aircraft based on twelve criteria. One of the most used MCDM techniques to select an acceptable al-

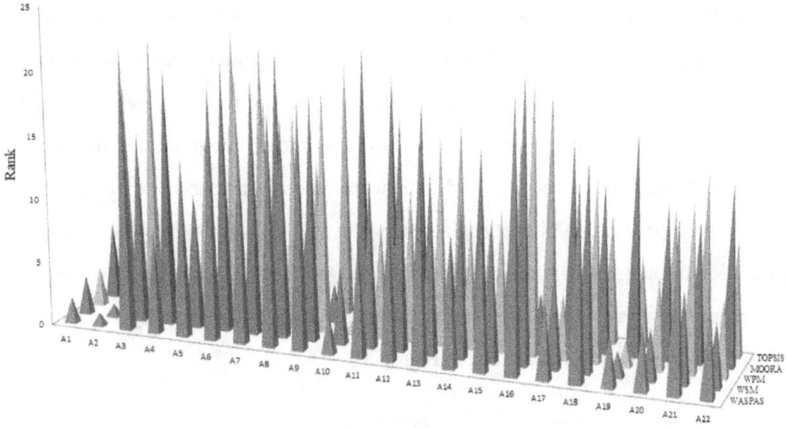

Figure 3.4: Rank analysis for various MCDM Approach

ternative was the TOPSIS method. The TOPSIS approach often provides less accuracy, leading to an imbalance in some practical scenarios. Therefore, this research developed a hybrid decision-making approach based on CRITIC and WASPAS techniques. A comparative analysis was performed between the proposed method and the CRITIC-TOPSIS method. As observed, the proposed method results in better accuracy than other methods when compared to the market statistics of midsize aircraft. Thus, the estimation of the best alternative selection via variants of criteria and its weights were dependent factors. In this investigation, the type of aircraft and aircraft configuration appears to be the prime objective criterion. Statistically, this hybrid method proved to be an efficient and accurate selection process. In a nutshell, this method is highly effective for investigating new design technology, infrastructure and operations measures, and economic effectiveness in the aerospace sector. Furthermore, to check the accuracy of this approach compared with MOORA, SAW, WPM, TOPSIS, and WSM methods demonstrated in Figure 3.4. This aircraft presents an opportunity and challenge to work in the design technology to emerge better fuel economy and improve economic competitiveness. It is also noticed that innovative technologies implemented on conventional aircraft with turbofans have a limited fuel economy. Instead, a wide-body airframe and its configuration is the revolutionary technology to reduce fuel burn, emit less carbon footprint, and improve performance. Nonetheless, the suggested framework focuses on gear-up other complicated issues in aerospace and industrial engineering. The next chapter provides the concept and application of the most Multidisciplinary Design Optimisation Techniques to be adapted to account for various design constraints and performance objectives by regulatory and industry requirements are illustrated in Figure 3.5.

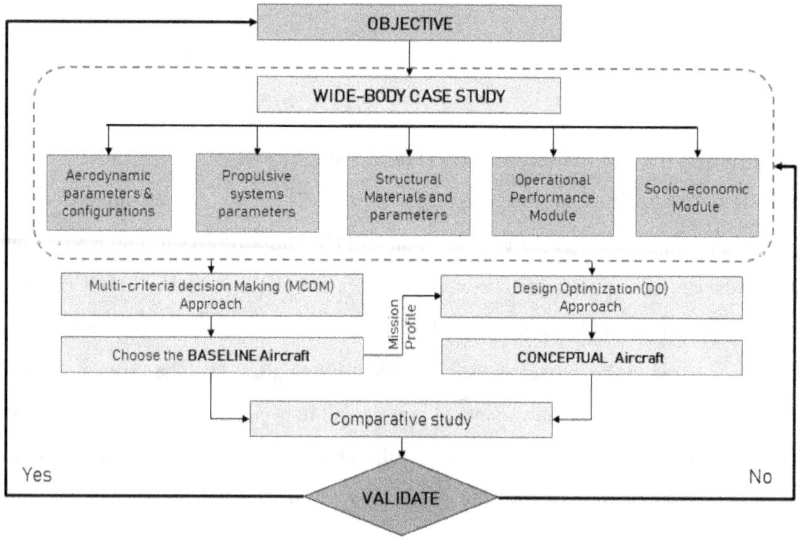

Figure 3.5: Future scope

Chapter 4

Multidisciplinary Design Optimization Techniques of a Novel Fuselage Design

4.1 Overview: Design Optimization

Optimization challenges in numerous industrial contexts, as exemplified in the planning and development of aerospace systems provide a good illustration of such challenges [189]. With an increase in rapid design and system complexity, optimization techniques are to solve and characterize many problems. Optimization seeks to attain the "optimal" design concerning a specified set of prioritized criteria and constraints. These encompass objectives like enhancing productivity, strength, reliability, durability, performance, and other utilization and verification. As per the definition of "Design optimization" is an approach to design based on the mathematical understanding of a technical problem to aid in determining the best solution from a set of options. Such a design approach undergoes the following steps: (a) Problem Statement; (b) Objective function; (c) Design variables; (d) Design constraint and (e) Solver Approach.

Besides the design parameters, many researchers approached designing an efficient aircraft using design optimization techniques [111] [177], especially for industrial design are tabulated in Table 4.1. Design optimization techniques are an approachable method and drastically applied to transport aircraft, and unmanned aerial vehicles (UAV) in the last 6 years. Thus a detailed study of design optimization methods for transport aircraft is carried out by different researchers under different methodologies tabulated in Table 2. Singh et al., [[illegible]] discussed the ways to reduce MTOW by utilizing genetic algorithms optimization techniques. Furthermore, obtaining the objective using the same design constraint can be extended through other optimization techniques. Chai et al., [[illegible]] described how to enhance fuel efficiency using multi-point over single-point optimization techniques. Li et al., [[illegible]] applied the fidelity optimization technique to improve wide-body aircraft's competitiveness. Ma & Elham [[illegible]] designed a novel TF configuration using a surrogated model and compared it with the conventional configuration yields to improve 20% fuel economy and 7% MTOW. Liu and Jiang [15] represent the Design optimization of tail design to minimize the MTOW by 3.28% and provide insight into a further reduction that can be achieved using Boundary layer ingestion (BLI) and other active flow control (AFC) systems. The aviation industry has made remarkable achievements in boosting the economy of today's aircraft, particularly in lowering the fuel consumption of aircraft engines. NASA's Glenn Research Centre is researching propulsion technologies to improve the fuel economy of future airplanes, lowering total consumer costs and environmental effects [[illegible]][185]. Budziszewsk & Friedrichs [[illegible]] suggested the concept of BLI to be introduced, not only to reduce drag but to improve the power utility to the maximum. Moreover, the application of innovative analyses to examine the economic growth of these technologies, as well as life cycle analysis to assess the ecological footprint across all phases of the aircraft life cycle is critical for future research projects. The examination of different aircraft types and mission ranges will also provide insights into how the advantages of this technology are influenced by aircraft size and how system sensitivities change [[illegible]].

Within this section, the critical facts of the optimization process are established. Specifi-

Table 4.1: Research contribution on design optimization of transport aircraft

Authors	Significance Facts	Parametric study
[134]	Evaluating the systems performance using AHP	Variables: Price; Cost, Security, Reliability, service quality.
[165]	Initiates to improve fuel economy and transport aircraft design configuration using Genetic algorithm (GA).	Variables: range, cruise speed, wingroot chord, engine dry weight, and lift over drag ratio. Constraints: flight performance and design conditions.
[133]	A comparison of different techniques using FANP & FAHP and obtain the suitable result to select the best one.	Variables: Price, Cost, Security, Reliability, service quality.
[192]	Implementing Multi-point Optimization to improve the wide-body design to reduce fuel consumption and study through real-time flight missions.	Variables: flight performance and design conditions.
[90]	The implementation of MCDM in the Taiwan aviation sector such as airlines, airports, air traffic management.	Variables: Seat capacity, MTOW, Range, Cost, CASK, Comfort,
[11]	Presented a unique approach to obtain the design variables using MCDM. Further, utilized the variable to find the wide-body aircraft market competitiveness using Monte Carlo techniques.	Variables: economics, comfort, environmental impact, and adaptability. Constraints: Profitability, payload, emission, and flight performance.
[99]	Selecting best suitable commercial aircraft using Fuzzy APH & TOPSIS MCDM and to check the robustness of the solution.	Seat capacity, MTOW, Fuel consumption, LTO cycle, Range, Speed, Price, Fuel per seat, DOC.
[52]	Selection and evaluation process using Entropy & WASPAS MCDM.	Variables: Seat capacity, MTOW, Fuel consumption, LTO cycle, Range, Speed, DOC.
[201]	Hybrid Optimization approach was applied to overcome the aircraft weight and balance issues.	Constraints: flight envelops and design conditions.
[33]	By using Non-dominated Sorting Genetic Algorithm II (NSGA-II), controlled the influence flight mission weightage.	Constraints: design mission parameters of significance and. portraits the effect of off-design mission weightings on the designs

[ ]	Established a multi-objective optimization of the design parameters and compared it to two distinct optimized solutions such as (MLVM) More less-violations method (MOGA) Multi-objective genetic algorithm.	Variables: flight performance and design conditions.
Proposed Research	Implementing the concept of MCDM techniques to identify the baseline aircraft and compared with the novel design. Optimization techniques are used on the variable on the constraints parameters to fulfill the objectives.	Variables: Design parameters with aerodynamic efficiency Constraints: design mission parameters of significance and portraits the effect of off-design mission weightings on the designs

cally, the problem statement, flight mission profile, and constraints will be derived through the utilization of the proposed MCDM technique, which aids in effectively capturing the multifaceted aspects of the design problem. Furthermore, the aircraft mission profile will be meticulously chosen from the top-ranked and best-performing aircraft options elucidated in the previous chapter (Chapter 4), ensuring that the selected design aligns with the predefined objectives and criteria. The baseline aircraft (A310-200) has been determined to validate the result with a novel conceptual one in terms of retrofit design, structural contributions, or operational performances. Such indulging parametric changes and incorporating advanced technology to find the MTOW of the novel aircraft [ ]. Its solution can be either a global optimum or no feasible solution will exist. Finally, the results found in an optimized solution are reported and compared with the baseline discussed in this study. The robustness of the novel design was validated by comparing the operational performance of the baseline aircraft.

4.2 Flight mission Problem statement

Based on the proposed MCDM approach, an appropriate selection of baseline aircraft A310-200 was obtained, and collected the flight mission profile. A baseline aircraft with a seating capacity for 275 passengers to travel a distance of 6500 km, similar to the A310-200 flight mission, is discussed. An aircraft has a cruise speed of Macr = 0.79 at a cruise altitude of 12.2 km and a service ceiling of 43,100 feet. As per airworthiness standards, a necessity reserved range of 370 km along with an 8 percent reserved fuel and a 45-minute hold at 1,500 feet are adopted. A summary of the flight mission considered from engine warm-up to the landing phase is shown in Figure 2.4. The flight mission is confined to the top-level design requirement into three categories represented in Table 4.2. The present work is concerned with the design parameter selection and optimization of a

Table 4.2: Top-level design requirements

Requirement	Notations	Value	Units
Maximum Cruise Velocity	Vc	850	km/hr
Cruise altitude	h_a	12.2	km
Diversion Range	Rr	400	km
Seating capacity	n	275	-
Trapped fuel	-	6%	-
Engine	-	GEnx-1B54/P2	-
Thrust of each engine	T	255	kN
Specific Fuel Consumptions	SFC	0.37	hr^{-1}
Fuselage configuration	-	twin-aisle	-
Maximum take-off and landing field lengths	LTO	< 1600	m

transport aircraft to minimize the MTOW using a design optimization approach. Usually, the optimization design problem employs a set of design variables required to minimize (or maximize) objective function. These three elements—design goals, constraints, and design variables—are illustrated in Figure 4.1 and described in the following paragraphs:

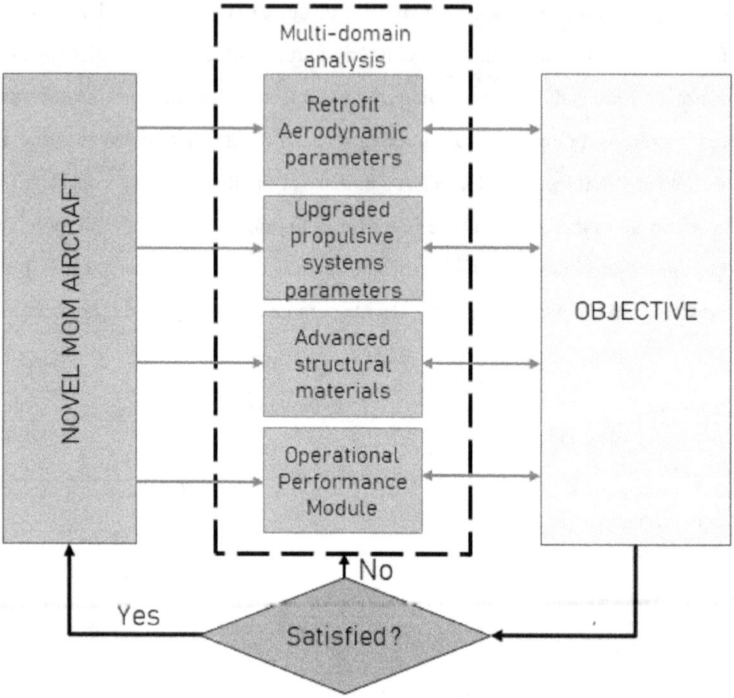

Figure 4.1: Flight mission based on the objective function

4.2.1 Objective function

The objective of this paper is to decrease the gross weight by design optimization of the middle-of-the-market widebody aircraft Airbus A310-200. The process starts with the data acquisition of the worldwide available widebody aircraft. As per the flight mission, the design sizing of a novel design is suggested to follow airworthiness standards FAR part 25. A novel MoM transport aircraft with a seating capacity for 275 passengers to

Table 4.3: Consideration of advanced technologies

Technologies	Assumptions
Advanced material and structures used	20% structural weight reduction
Upgraded efficient Engine	9.39 BPR with Engine OPR 35 High bypass ratio turbofan with TAPS

travel a distance as per the middle-of-the market and, integrated with advanced technologies to minimize MTOW to improve the performance of a novel aircraft. However, the predictions on the impact of modern technology on this aircraft are provided in Table 4.3.

4.2.2 Weight Module

Additionally, the constant parameters obtained from the baseline aircraft are represented in Table 4.5, and the objective model is developed in Equation (2.2.2). Thus, to minimize the objective function was stated as
Min MTOW = Minimize FF (X) = W_{TO}.
w.r.t, wing, fuselage, horizontal tail, vertical tail
s.t design constraints.

4.2.3 Design variables

As the conceptual design is concerned, the design variables (X) are obtained accordingly. There are fifteen design considered variables such as wing area (S_w), wing span (b_w), wing taper ratio (λ_w), fuselage length (L_f), fuselage height (h_f), fuselage upsweep angle ($\theta_{upsweep}$), no. of abreast ($N_{abreast}$), horizontal tail surface area (S_{HT}), horizontal tail span length (b_{HT}), horizontal tail taper ratio (λ_{HT}), vertical tail surface area (S_{VT}), vertical tail span length (b_{VT}), vertical tail taper ratio (λ_{VT}), Range (R) and aerodynamic efficiency

(L/D) and each expression are explained in previous section. The design geometry of other parameters has been assumed to be a similar wing-tail configuration of the Airbus A310-200, and the engines have a wing-mounted arrangement. This study presumes to use of advanced lightweight materials in the airframes. Hence, the mathematical function of MTOW is

X = function (S_w, b_w, λ_w, L_f, h_f, $\theta_{upsweep}$, $N_{abreast}$, S_{HT}, b_{HT}, λ_{HT}, S_{VT}, b_{VT}, λ_{VT}, R, L/D).

In addition, the sizing of the design parameter should be pursued based on the objective function, the 'dedicated-to-a-type' design illustrated in Table 4.5, and detailed code programmed at MATLAB R2014a are available in Appendix. The upper and lower bounds of design variables were determined and chosen based on the statistical dataset.

4.2.4 Design constraints

The most crucial part is to formulate an acceptable set of constraints for a certain design process. The constraint function contains a set of equations that are presented in the form of upper and lower-limit equations. There are two types of constraints present in the Constraints function they are Equality contains and Inequality constraints. In this research, 15 constraints are considered to satisfy the aircraft performance as per airworthiness standards and marked in Table 4.4. The Inequality constraint returns a negative value when the values were assigned that represent the feasible solution possible at that point. If any of the constraints have a positive output value that means the value is out of bounds and causes infeasibility at that point. The main objective of creating a constraint function is to provide limitations for the process of optimization. [[illegible]]

Table 4.4: Constraint requirements as per airworthiness

Design Constraint		Equation	Limits	Reference
	AR_W	(2.2.8)	$9.9 < AR_W < 6.9$	[116] [13]
	FR	(2.2.7)	12.3 < FR < 8.2	[116] [103]
	AR_{HT}	(2.2.8)	3 < ARHT < 5	[116]
Retrofit design	AR_{VT}	(2.2.8)	1.3 < ARVT < 2.0	[116]
	S_{TO}	(2.2.23)	$1400 < S_{TO} < 2550$	[165] [43]
Operational performance	ROC	(2.2.24)	5.08 < ROC < 28.9	[165] [152]
	S_{LO}	(2.2.25)	$1300 < S_{LO} < 2100$	[165] [43]
	$C_{D,0}$	(2.2.13)	$5.08 < C_{D,0} < 28.9$	[76]
	W_E/W_{TO}	(2.2.3)	$0.3 < W_E/W_{TO} < 0.7$	[165] [43] [144]
Structural integrity	V_f	(2.2.11)	62 < Vft < 85	[165] [76]
	$\overline{V_{HT}}$	(2.2.27)	$\overline{V_{HT}} = 1$	[172] [154]
	$\overline{V_{VT}}$	(2.2.28)	$\overline{V_{VT}} = 0.07$	[172] [154]
Static Stability characteristics	ζ_{ph}	(2.2.30)	$\zeta_{ph} > 0.05$ or 0	[154] [16] [184]
Egronomics	D_f	(2.2.32)	$5 < D_f < 6$	[153]
	N_{pitch}	(2.2.32)	$5 < D_f < 6$	[153]

4.2.5 Solver

The design optimization of the baseline aircraft Airbus A310-200 has been done by using the MATLAB optimization toolbox. This is a feature provided by the MATLAB application to analyze a linear or non-linear function through a set of constraints to provide an optimized value for one or more variables that are provided in the main function. This toolbox follows either a maximization or minimization approach. The main function/-function problem is the function that is solved through the iteration process by the selected solver. Based on the type of the function problem like Linear, Non-linear, Quadratic, and second-order cone programming functions. There are several types of solvers were present in the MATLAB Optimization toolbox. Each solver contains a particular internal functionality to solve the respective problem. These internal functionalities are called optimization options, pre-defined functions that help to perform operations like recognizing the type of problem for the solver, plotting the obtained data as histograms, curves, etc..., providing order of optimality, and for other operations to personalize the data output. The optimization solver requires upper and lower bounds to perform the optimization process. It allows the user to give an initial point ie; to specify the value within the limits or bounds to initialize the optimization process from that point.
The considered objective function is to improve the performance of the aircraft hence the optimization problem in this paper is a constrained minimization problem. FMINCON is a MATLAB solver used to solve constrained minimization problems. 'Fmincon' solver can be assigned to the problem using the 'Optimoptions' function. The upper and lower limits within which the optimization takes place for the function variables. The initial point should be provided to start the optimization. This allows to initiation of the process from any point within the provided limits. The process followed to solve the optimization problem of this paper is a non-linear constrained minimization. To enhance the aircraft design, a set of design variables with a limited constraint helps to examine the value of

MTOW. Such a framework can create an improved design to fulfill the objective using a design optimization approach. Several researchers handled different design optimization approaches to solve effective design models. Many authors followed the genetic algorithm (GA) to optimize the required objective [143] [151]. Merely, GA can obtain global solutions and results to converge local optimum value. Thus, the current study prefers the interior point optimization technique over other techniques due to its solver algorithms and obtaining a globally optimized solution. In addition to it, the implemented approach has a high computational time, unlike GA. The robustness of this approach solves all kinds of problems such as linear and non-linear. The optimization was carried out on the GNA University workstation with a 4.1 GHz CPU, i7 12th generation processor, and 16 GB of RAM utilizing MATLAB version R2014a to estimate the objective function. Based on the considered input values, the objective MTOW value was obtained and tabulated in Table 4.5. To validate the result, the outcomes are compared with the parameters of the Airbus A310-200 in Table 4.6.

4.3 Results and Discussions

This section presents the outcome of the objective function of a proposed optimization approach. In addition to it, advanced technology has been assumed to be incorporated into the baseline model, are noted in Table 4.3. Based on the domain considered, this chapter focuses on implementing interior point optimization to minimize the MTOW. By using this optimization technique, not only is the MTOW of the aircraft minimized but also to improve performance such as take-off distance, Rate of climb, and aerodynamic efficiency. This function presented two cases of optimization: case 1 by only fuselage optimization and case 2 by changing wing-fuselage-empennage optimization. The optimized results converged after 231 iterations as shown in Figure 4.2. After the convergence was made, the optimized values were obtained in the comment section. The optimized parameters provide the layout configuration of our proposed model as indicated in Fig 4.3.

Furthermore, the study extends the better influence on performance for both cases 1 and 2 are illustrated in Figure 4.4.

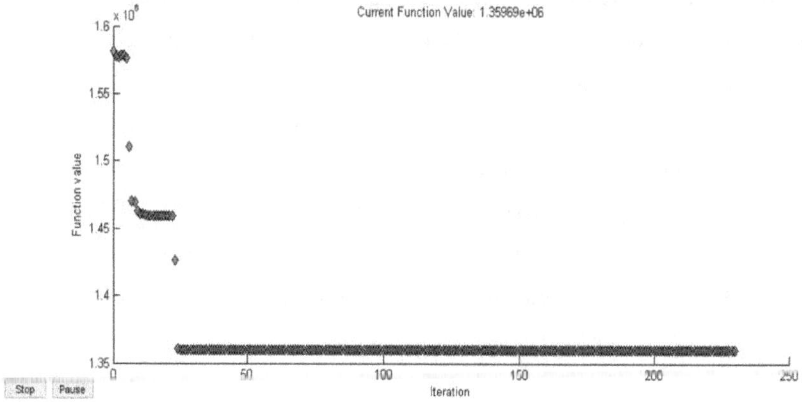

Figure 4.2: Convergence history of objective function

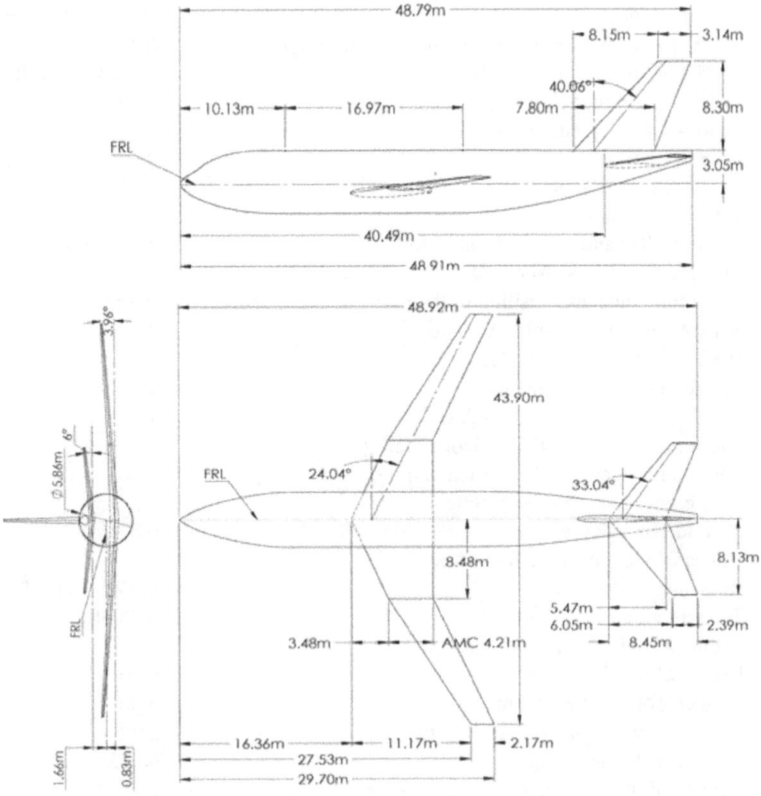

Figure 4.3: Optimized novel aircraft design configuration

Table 4.5: List of constant parameters

Parameters	Specifications
Density, $\rho(kg/m^3)$	0.01841
Dynamic viscosity, $\mu(Ns/m^2)$	1.475
Wing configuration	Low wing
Wing planform	Backward sweep
Thickness to chord ratio of Root of wing, $(t/c)_{w,R}$	0.12
Thickness to chord ratio of Tip of Wing, $(t/c)_{w.T}$	NASA SC 0610
t/c location of root chord, k_R	0.37
t/c location of tip chord, k_T	0.38
Dihedral angle	2.3°
Fuselage configuration	Twin aisle
Tail configuration	Conventional
Stabilizer Tip and Root section airfoil	NASA 0009
Maximum thickness-to-chord ratio, $(t/c)_{max,HT}$,	0.09
Maximum thickness position of the horizontal tail, $(x/c)_{m,HT}$	0.307
Vertical tail Sweep angle, $\Lambda_{C/4,VT}$	44
Horizontal tail Sweep angle, $\Lambda_{C/4,HT}$	34
Interference factor of wing, Q_w	1
Interference factor of the fuselage, Q_f	1
Interference factor of the horizontal tail, Q_{ht}	1,04
Interference factor of the vertical tail, Q_{vt}	1.04
Interference factor of the nacelle, Q_n	1.5
Oswald efficiency, e	0.8
Universal Gas constant, R (J/kg/K)	286.9
Roughness factor, K_r	$6.33956 * 10^{-6}$
Tail arm moment, l_{ht} (m)	0.55 of L_f
Ultimate load factor, N_{ult}	6
Length of nacelle, L_n (m)	6.06
Diameter of nacelle, D_n (m)	3.06
Density of wing material, $\rho_{mat,W}(kg/m^3)$	2106
Wing density factor, $K_{\rho,w}$	0.0025
Density of fuselage material, $\rho_{mat,f}(kg/m^3)$	2682.3
Fuselage density factor, $K_{\rho,f}$,	0.0025
Elevator-to-tail chord ratio, C_e/C_T	0.35
Density of HT material, $\rho_{mat,HT}(kg/m^3)$	2106
HT density factor, $K_{\rho,HT}$	0.02
Landing gear configuration	Multi-bogey
LG Tires	2; 8
Retractable landing gear factor, K_{ret}	1.07
LG density factor, K_{LG}	0.32
Landing gear height, H_{LG}	0.5
Engine weight factor, K_E	3
Number of engines installed, N_E	2
Thrust, T (kN)	255.3

Table 4.6: Optimized outcome of the proposed model

Design variables	**Baseline**	**Generic Value**		**Optimized Value**	
		Lower bound	**Upper bound**	**Case 1**	**Case 2**
Sw	219	122	845	219	200
bw	43.9	33.9	79.8	43.9	42.87
λ_w	0.259	0.2	0.35	0.259	0.32
L_f	46.70	33.8	76.73	48.91	47.2
h_f	5.7	4	6.2	5.72	5.72
$\theta_{upsweep}$	17.2	16	20	17.59	17.3
$N_{abreast}$	8	6	8	8	9
S_{HT}	76.83	63	81.50	76.83	69.10
b_{HT}	16.26	12.45	31.29	16.26	16
λ_{HT}	0.34	0.25	0.38	0.34	0.385
S_{VT}	45.24	30.5	45.5	45.24	52.2
b_{VT}	8.3	7.5	9.5	8.3	7.97
λ_{VT}	0.39	0.30	0.32	0.39	0.31
R	6500	5500	9500	7905	9014
L/D	16.3	15	20	16.7	17.12
Design Objective (N) MTOW	14,12,640	-	-	13,59,690	13,44,250

The overview of the proposed model is indicated with the red color zone, whereas the baseline model is highlighted using green color. From the perspective of a side view of the aircraft it clearly shows the difference in the streamlined body postulated as per optimized value. It is noticed that the rear portion of the fuselage is highly cambered and slight increase in the fuselage length. The parametric design has been created using Solid Works design software through the loft, and spline command tools. The indications of the design parameters are discussed with the following consequences:

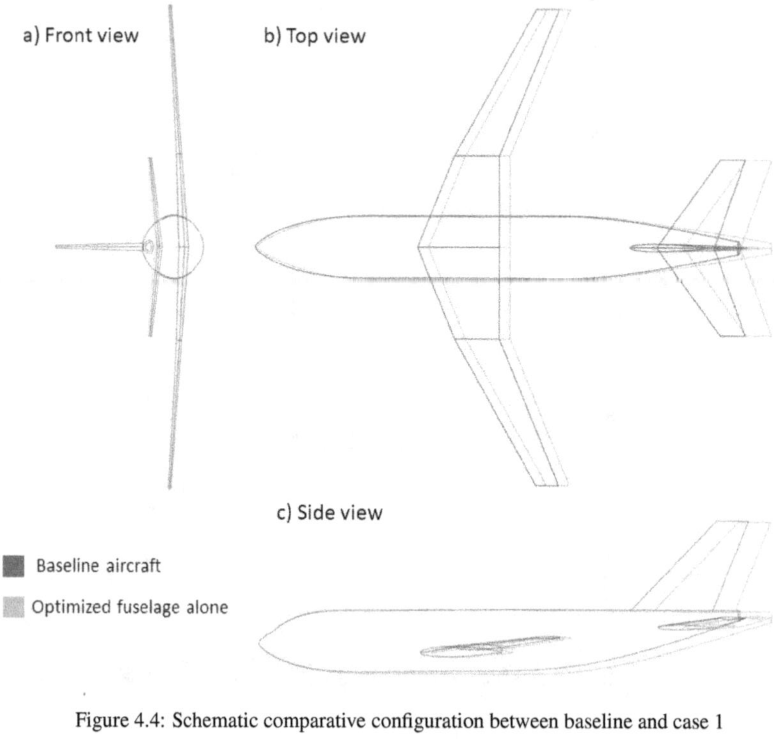

Figure 4.4: Schematic comparative configuration between baseline and case 1

a. The value of wing area and wing span has reduced to 200 m^2 and 42.87 m, respectively. As a result, an increase in the Aspect ratio (AR) from 8.8 to 9.2 has been noticed. Due to its impact, there is a possible way to improve aircraft performance. Further, an increase in the taper ratio from 0.26 to 0.32 has been adopted. Aerodynamically, this value will help to reduce the induced drag. Hence a drastic change in the taper ratio has been indicated.

b. The value of fuselage length and diameter both increased to 48.8.2 m and 5.86 m, respectively. As a result, a slight decrease in the fineness ratio (FR) from 8.3 to 10 has been noticed. The increase in the fineness ratio up to the optimized value will favor the form drag factor. The upsweep angle and fuselage height are having negotiable changes.

c. The value of the Horizontal tail (HT) wing area and wing span has reduced to 69.1 $m^2$2 and 16 m, respectively. As a result, an increase in the HT Aspect ratio (AR) from 3.4 to 3.7 has been identified. Similarly, the taper ratio leads to an increase in the value from 0.34 to 0.385.

d. The value of Vertical tail (VT) wing area and wing span has increased to 52.2 m^2 and reduced to 7.97 m, respectively. As a result, a reduction in the VT Aspect ratio (AR) from 1.64 to 1.22 has been predicted. Further, the taper ratio leads to an increase in the value from 0.3 to 0.31 has been obtained.

e. A drastic improvement in aerodynamic efficiency called lift-to-drag ratio has been noticed and the value increased from 16.3 to 16.7. As a result, the parasite drag coefficient has reduced the value from 0.0207 to 0.0197. The reduction of the drag coefficient has caused due to the parametric design change and incorporation of advanced technologies. Hence, it is proven that this optimization result shows the reduction in MTOW of a novel aircraft as shown in Figure 4.5.

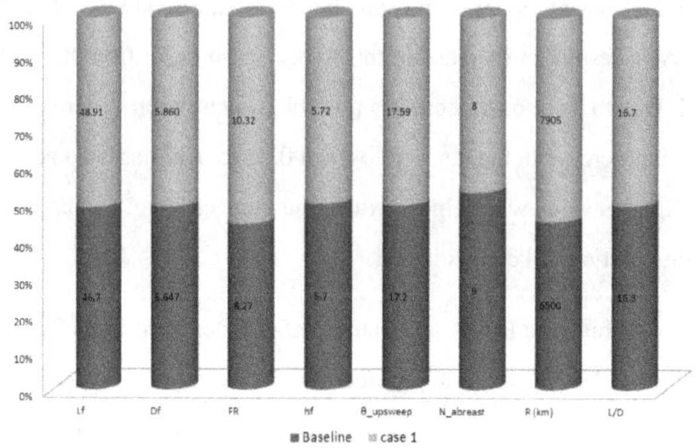

Figure 4.5: Comparative study between baseline aircraft and case 1

Comparative Studies with baseline

To understand the robustness of the design, it splits into two cases: with and without the advanced technologies mentioned in Table 4.3. The first case is to investigate a novel fuselage design by applying assumed technologies in the solver, and the second case is to investigate a novel aircraft design by applying assumed technologies in the solver and compare both cases with the baseline aircraft model Airbus A310-200. In the first case, no changes in wing, tail, and horizontal tail parameters, with, a 4.73% increase in the fuselage, and a 6.03 % increase in diameter which resulted in further improvement in fineness ratio by 24.79% compared to the baseline design. In this case one also analyses, the upsweep angle of the fuselage which increases by about 2.27 %, and 11.11% decrease in the number of abreast of the novel aircraft cabin. Furthermore, a hike in range R is obtained at about 21.62% and an aerodynamic efficiency leads to an increase of 2.45%. The outcome of the optimization shows aerodynamic efficiency has high fidelity due to its design threshold value of considered parameters. Aerodynamically, the fineness of the fuselage ranges from 8 to 12 at a certain Reynolds number leading to improved parasite drag of

an aircraft and especially, form drag factor. The aerodynamic efficiency of the optimized value seems to have greater value as compared with baseline aircraft. Additionally, advanced structural material and an upgraded engine initiate the reduction of the structural weight of the aircraft, hence creating a low drag profile, fuel economy, and emission as well. The concept of fuselage shape is considered an advanced technology. Therefore, the flow over a fuselage body is optimized to be streamlined to reduce drag. Additionally, the parasite drag coefficient ($C_{D,0}$) is also evaluated to predict the improvement in the novel aircraft which shows a 10% reduction of drag as compared with the A310-200 model. Figure 4.5 illustrates the overall improvement of aircraft design parameters and results to minimize the MTOW. The use of the optimization approach resulted in an overall improvement in the MTOW value of around 3.75%.

As previously mentioned equation (2.2.2) shows that factors affect the MTOW of an aircraft. Furthermore, the analysis of this model has been performed. The outcome of this optimization results shows an improved value of the empty weight, fuel weight, and performance factors that have been noticed in Fig 5.2b. The first factor is the empty weight which significantly depends on the design parameters of the aircraft's component and is marked in equation (2.2.4). The obtained model yields the empty weight fraction to 0.42 by implementing advanced material into the airframe and component installation factor. Therefore, a drastic decrease in empty weight fraction was around 25.59% as compared to the baseline aircraft Airbus A310-200. A significant change in empty weight has a substantial impact on design parameters and mathematical relations are expressed in equation (2.2.4d - 2.2.4g). A notable improvement in W_E and $\frac{W_E}{W_{TO}}$ is due to the application of lightweight structures, materials used in the structure, and changes in design parameters as per the design objective. Looking into the fuel weight factor, which depends on the mission profile (i.e., Range and flight velocity) along with the engine configuration, and aerodynamic efficiency are expressed in equation (2.2.6). As per the considered optimization model, 27% and 32% have been increased in fuel weight and fuel weight fraction compared to the baseline, respectively. There is a possibility for an extra capacity of fuel to endure better range compared to baseline aircraft. Hence, improved aerodynamic efficiency and specific fuel consumption for the required mission could clarify a slight increase in fuel weight and fraction. Thirdly, to study the performance factor of the aircraft through take-off field distance (S_{TO}). Theoretically, S_{TO} is the sum of ground distance (S_G), rotational distance (S_R), and transition distance (S_{TR}) by keeping the constant value of take-off velocity(V_{TO}), rotation time (T_R), stall velocity (V_{stall}), load factor during take-off, and radius of turn (R), respectively. Hence, the optimized TLO value turns out to be around 1461 km and improved by 8.063% compared to the baseline. Likewise, numerous parametric analyses can be performed to find the robustness of this optimized result.

4.4 Summary of key findings

The highlights of this section present a novel conceptual design of MoM transport aircraft [ ] [ ] [ ] [ ]. Nowadays, researchers reported that the MoM aircraft leads a streamlined progression in transportation and tourism. The motivation of this paper is to minimize the gross aircraft weight (MTOW) using an optimization approach. The main feature of novel transport aircraft is a low wing configuration with advanced lightweight materials, an efficient engine, and an improved retrofit design. The results show that implementing advanced technology and design parameters triggered to decrease in the MTOW. Additionally, implementing efficient twin GEnX engines boosts performance, improves fuel economy, and reduces emissions. The influence of aerodynamic parameters on the wing, fuselage, and empennage has affected the MTOW value. As a result, the aerodynamic characteristics of wing planform geometry, fuselage geometry, and empennage planform geometry are vital for aircraft design. The change of these geometrical characteristics correctly, concerning their impacts on the aircraft's aerodynamic parameters, can enhance the performance. Numerous engineers aim to optimize the fuselage fineness ratio to decrease the fuselage drag. This optimization approach obtained a fineness ratio of 10.32 and was comparatively reduced by 1.81%. The design variables directly connected with the lift-to-drag ratio (L/D) have been estimated and the result is improved by 24.79%, thus achieving better aerodynamic efficiency compared to the baseline aircraft. On the other hand, a preliminary performance study of a novel aircraft has been performed. This study targeted to find the impact of empty weight, fuel weight, and performance parameters that affect the design. However, the parametric studies reported that there is a significant potential to improve the performance in terms of structural weight, fuel weight, and take-off distance which were reduced by 25%, increased by 25%, and reduced by 8.063%, respectively. Implementation of novel transport aircraft will potentially lower the drag, especially after embedding the concept of advanced technologies. Therefore, the conclusion shows that the novel design with the assumed technologies triggered the environmental and economic pillars of sustainability. The proposed aircraft design obtained from the optimization modeling improved the MTOW compared to the baseline aircraft. Moreover, the empty weight, the fuel weight, and the take-off distance were improved which reveals better structural integrity, fuel consumption, and aircraft performance.

The article states the prospects and extension of the work in two ways. Firstly, the robustness of the optimization model can be compared using parametric analysis. In continuation to this, comparison can be shown among the combination of technologies with one another or adding upgraded retrofit designs. Secondly, both computational and experimental work can be carried out to characterize the performance of the flight. After successful flight testing, the certification of novel aircraft can proceed for better airline profitability and fuel economy. Thirdly, as indicated in the research study, different multidisciplinary design optimization models can be performed by other parameters. Besides, the difficulties of acquiring data, certain parameters are evaluated, and the interactive parametric study is not taken into account [165] [49] [57]. As a result, more analysis of the interaction between the parameters depending on specific application situations can be recommended for further study. In future research, the proposed model can be merged with favorable hypotheses to increase performance efficiency and result accuracy. This research may be expanded further with an emphasis on a different approach for validation through computing time to solve the objective functions in addition to the inputs.

Chapter 5

Sensitivity Analysis of Design Optimization Technique for the novel aircraft design

5.1 Background

In this context, sensitivity analysis is a great technique that allows us to dive into the complexities of optimum solutions and understand how differences in basic inputs or assumptions may have a substantial influence on the outcomes. In the realm of research, it is imperative for any investigator to validate the model and scrutinize its design parameters rigorously. The model under scrutiny must not only be verified for its superiority when compared to existing models but also undergo a comprehensive examination of the aircraft design variables through sensitivity analysis, employing both relative improvement and regression methods. The primary objective of this chapter is to thoroughly assess the model's robustness because, in the context of aircraft design, the enhancement of performance and increased efficiency are of paramount importance. Given the inherent complexities associated with aircraft design and its multitude of parametric variables, a meticulous evaluation of the objective model becomes indispensable. This evaluation must be carried out with a keen eye on the design parameters, meticulously comparing them with the existing body of research literature to ascertain their relevance and reliability.

Many researchers worked on different approaches and schemes to analyze the robustness of different objective goals such as ANOVA, regression, "derivative-based on local methods", or other hybrid meta-analysis methods[112] [175] [94]. Several researchers and scientists reported that the fuselage creates an ample drag and needs to be reduced through emerging technologies. The basic studies said that the construction of the fuselage varies based on the aircraft design and its purpose. NASA demonstrated a concept of double bubble D8 wide-body fuselage to minimize the MTOW and drag force at Mach 0.74. Furthermore, investigation results generated a 29% reduction in fuel burn and cumulative noise emission as well. Yutko et al., [194] implemented the concept of boundary layer ingestion (BLI) at the fuselage's tail section to maintain the laminar flow at the exit wake of the aircraft. A student's project from Tu Delft has presented an innovative V-fuselage design to improve passenger comfort without degrading the aircraft's performance. Chai et al., [illegible] worked on the multi-point optimization approach to improve the fuel economy and cost optimization of a wide-body aircraft and discussed the modifications based on the objective function. Thu et al., [3] discussed a modified aircraft design to reduce the wake of an aircraft using a sequential quadratic programming algorithm (SQP). Chavan and Pawar [illegible] reported improved performance by optimizing the shape of the fuselage design. Recently, Unal et al., [5] demonstrated different aircraft analysis software such as piano, Advanced aircraft analysis software, and other real codes with Python and MATLAB software.

In the previous chapter, the focus was on optimizing the Maximum Takeoff Weight (MTOW) of the aircraft, and this optimization was achieved by considering parasite drag consideration as the main objective function. This approach led to a reduction in drag and an overall improvement in the aircraft's performance. However, to ensure the validity and reliability of these results, it is essential to refine the model further. This refinement involves shifting the primary focus from MTOW to another critical parameter: the drag coefficient to check the robustness of the proposed design concept. Just as with any conceptual design in engineering, the process of validation is crucial. By shifting the focus to the drag coefficient and conducting both computational and experimental assessments, the novel work can assess a more comprehensive and robust understanding of the aircraft's performance, thereby enhancing confidence in the design's effectiveness and reliability. This approach ensures that the design decisions are not only theoretically sound but also practically proven through rigorous validation procedures. The chapter is to provide an answer as follows:

(a). How to compute the baseline and novel aircraft design and its performance parameter?

(b). What are the key findings and managerial insights for the achieved goal?

The answer to the following questions is provided in the consecutive sections.

5.2 Sensitivity Analysis

In the context of operations research, sensitivity analysis is a great technique that allows us to dive into the complexities of optimum solutions and understand how differences in basic inputs or assumptions may have a substantial influence on the outcomes. The method used in this article reveals how changes to inputs, definitions, or modeling can improve the accuracy and resilience of our conclusions as basic factors change. The aim is to analyze the design variable utilized to solve the objective function as expressed as equation 2.2.2. This chapter is primarily focused on conducting a comprehensive comparative analysis of two distinct objectives within this pursuit - the reduction of Maximum Takeoff Weight (MTOW) for two different cases. The central question here revolves around how a change in the objective function influences critical aspects such as structural integrity, aerodynamic performance, and passenger comfort for change in design variables and constraints. The comparative evaluation is essential to discern which of the two objectives demonstrates superiority, and the selected superior objective will subsequently undergo a rigorous validation process. This validation process is integral to ensuring the reliability and effectiveness of the chosen approach in addressing the shared goal of drag reduction, which is of great significance to the future aviation industry.

5.2.1 Design calibration Tool

Many researchers and scientists have developed different aircraft design calibrating analysis tools to design the aircraft according to the design requirement. Some design tools are fair and good for certain aircraft's component analysis and some are not. There are some popular design tools are drastically used by researchers and scientists such as Aircraft Synthesis Program (ACSYNT) [[illegible]] [[illegible]], Advanced Aircraft Analysis Program (AAA) [[illegible]] [[illegible]] [[illegible]], PIANO [[illegible]], a web-based Aircraft Design Pad (ADP) [[illegible]], and RDS-Student [[illegible]] [[illegible]]. After careful study of merits and demerits, AAA 5.0 under DARCorp shows an overall better response on transport aircraft design [[illegible]]. This chapter highlights the utility of AAA 5.0 version software to calibrate aircraft performance.

Advanced Aircraft Analysis 5.0

In this chapter, the analysis of the considered aircraft's performance parameters was carried out using the AAA 5.0 version software tool. To ensure the successful execution of the design process and the attainment of the desired performance metrics, a stepwise procedure, as outlined in the AAA manual, was meticulously followed. The process commences with a focus on aircraft certification and type categorization, where the specific regulatory and operational classifications are established, providing a fundamental framework for subsequent design decisions. Following this, the attention shifts towards the meticulous selection of various aircraft components and their configurations, a pivotal step that sets the stage for the entire design process. Moving forward, the subsequent phase involves the careful selection of the payload for the aircraft, a decision that significantly influences the aircraft's overall performance characteristics. Each of these steps is integral to the overall design process and necessitates thorough consideration and adherence to regulatory and operational standards. Gradually, in this sequential process is the weight estimation and performance prediction stage, a crucial aspect of aircraft design. Here, the tool calculates the expected weight of the aircraft, accounting for all its components, including payload, fuel, and structural elements by predicting the aircraft's performance metrics as shown in Figure 5.1. This versatile software serves as an invaluable resource, enabling the prediction and calculation of a wide array of design variables and parameters essential for a comprehensive aircraft design process. The AAA design tool, renowned for its user-friendly interface and efficiency, simplifies and streamlines the complex calculations required for the design of aircraft. In pursuit of a successful research endeavor, GNA University has thoughtfully embraced the AAA design tool as a key enabler. By making this software accessible and readily available within the university premises to excellence in this research work.

Initial Calculations

Based on the objective functions, the unknown variable values have been obtained as discussed in the previous chapter. Among the unknown variables, certain other parameters are required to find inorder to substitute the values into the AAA 5.0 software. Before starting the initial estimation of the parameters and performance of the aircraft, first need

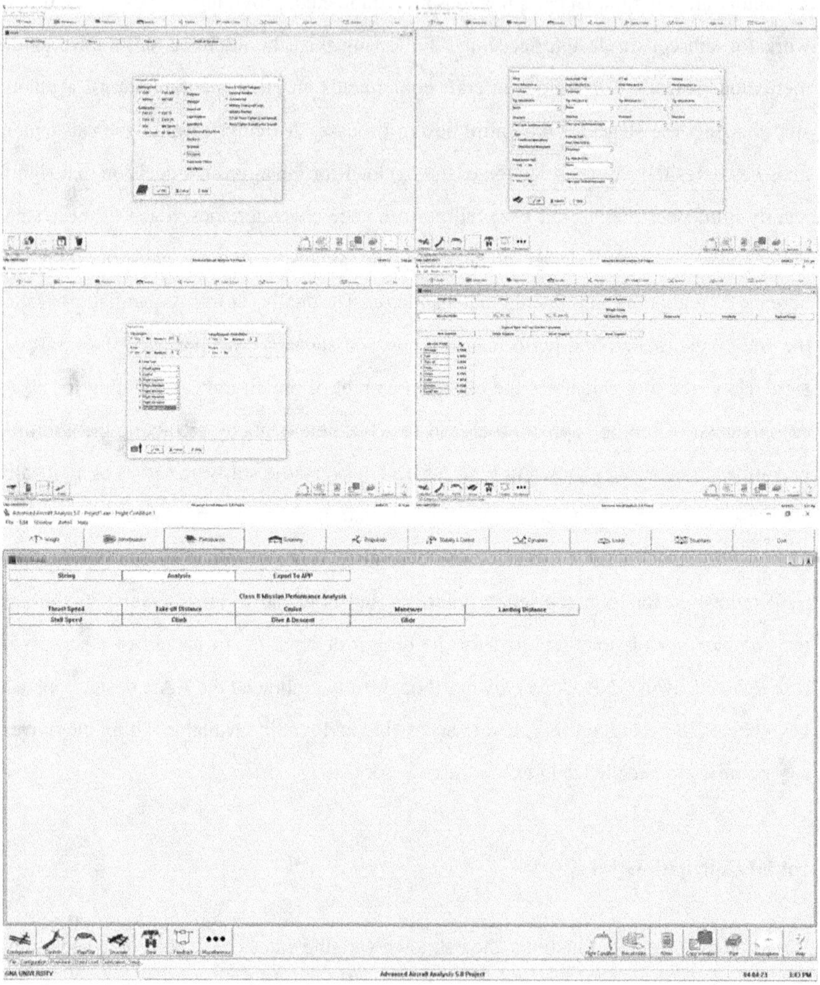

Figure 5.1: Some input values in AAA 5.0

to feed the flight conditions, design configurations, and type of classification of the aircraft as shown in figure 5.1. The configuration setup toolbar consists of configuration, controls, flap or slat, structure, engine, nacelle, duct, landing gear, wing storage, or fuel tank, and payload or crew. Each of the toolbars has its individual dialog box, to feed the configuration or values for each tab as per the aircraft design concepts. Any aircraft needs to follow the certification and the type of flight, this AAA 5.0 software provides a certification toolbar to select the certification and classification of flight. According to this research, the selected aircraft model type is FAR 25 (Transport, Civil) and is categorized under class III as per the instructions provided in the AAA 5.0 manual. Once the selection of flight conditions is done, the next is to set up the unit and calculation method to be selected for the further procedure of the rest of the module calculations.
The first module is the weight module or weight sizing of aircraft. To proceed further, the first step is to calculate the diameter of the fuselage using the equation (2.2.32). Applying the dimensions of the aisle width, seat width, number of aisles, number of abreasts, and clearance thickness as per the standard design protocol, to obtain the value of the diameter of the fuselage. Then, using the equation (2.2.7), estimate the value of fineness value of the fuselage. The next step is to estimate the structural weight of the fuselage using the equation (2.2.4b), by applying the constraints and constants values taken into consideration as illustrated in Table 4.5. The above calculation will help to obtain the empty weight and empty weight fraction using the equation (2.2.3) and (2.2.4) for the complete mission profile as shown in figure 2.4. After calculating the weight value, the class I and II weight modules are calculated followed by weight fraction and calculating aircraft component empty weight using equations (2.2.4).

5.3 Results and Key findings

This session discusses the design and performance output obtained from the AAA 5.0 software. Initially, the aircraft class settings and FAR part ought to be selected as CAR 25. As per the considered constraints and constant values, the MTOW value is estimated at initial iterations. Stepwise, applying the weight fractions and other aircraft details dimensions and parametric values, the unknown dimensions and performance parameters were obtained.

5.3.1 Comparative studies: Case 1 and Case 2

Both instances have been compared to assess the model's robustness based on performance, aerodynamics, and structural integrity. Initially, a geometric relative change is studied for the considered cases. It is divided into two cases—one without and one with the wing-empennage optimization to understand the resilience of the design are studied in detail. Case 1: In case 1, the study was solely carried out on the fuselage optimization, and other aircraft components are fixed. Initially, the fuselage design is optimized to reduce the non-lifting parasite drag by 10%. Simultaneously fig shows the effect of the modified fuselage on the baseline aircraft. It has been significantly noted that the drastic increase in fuselage length (L_f) and fuselage diameter (D_f) is 48.91 m and 5.86 m, respectively. While considering the other aircraft components kept unchanged. Due to its random increase in L_f and D_f , which affected the design parameter called fineness ratio. In comparison to the baseline, the abrupt increase in fineness ratio to 8.35 which hiked by 0. 8%. This component is mathematically tied to the parasite drag coefficient, as given in the equation (2.2.13). As a result, increasing the fineness ratio tends to decrease the drag coefficient or drag force. However, optimizing the fuselage alone may be able to improve empty weight, empty weight fraction, ground run, and phugoid damping. In contrast, such a strategy increases fuel weight and MTOW by 27% and decreases by 25%, respectively. Nonetheless, an increase in fuel weight indicates an increase in aircraft range compared to the operating range, which has increased MTOW by 21.62%. Additionally, the change in design parameter also affected the lift-to-drag ratio (L/D) which also improved about 2.45% compared to baseline flight.

Case 2: In this instance, this optimization technique takes into account the majority of the primary components, including the fuselage, wings, and empennage. The remaining design parameters and constraints, however, are maintained at their initial setting as the baseline aircraft. Considering all twelve different design variables such as fuselage length, diameter, wing area, wing span, wing taper ratio, wing sweep angle, horizontal tail area, horizontal tail span, horizontal tail taper ratio, Vertical tail area, Vertical tail span, and Vertical tail taper ratio. It has been noticed that fuselage length, horizontal span length, and vertical span length dimension remain unchanged compared to baseline aircraft. Simultaneously, a notable increase in fuselage diameter, wing area, wing span, wing taper ratio, wing sweep angle, horizontal tail area, horizontal tail taper ratio, and Vertical tail area are 5.98 m, 223.70 m^2, 46.34 m, 0.35, 57.98 m^2, and 33.80 m^2, respectively. Even though a few parameter values have decreased in the horizontal tail taper ratio, and the vertical tail taper ratio are 0.25 and 0.28, respectively. Consequently, an increase in the wing tip chord and aspect ratio, whereas, a decrease in the wing root chord, horizontal aspect ratio, vertical aspect ratio, and fineness ratio. The overall changes in the geometric values are exemplified in Fig 5.2a.

Analyzing the based on the geometric parameter, the fineness ratio of the fuselage has increased in the aforementioned scenarios by 4.73% and 3.901%, respectively. Case 1's fuselage design modification has been optimized to meet the specified constraints. In this scenario, the fuselage design is only modified through optimization techniques resulting in only a change in the fineness ratio. In contrast, scenario 2's wing to empennage design

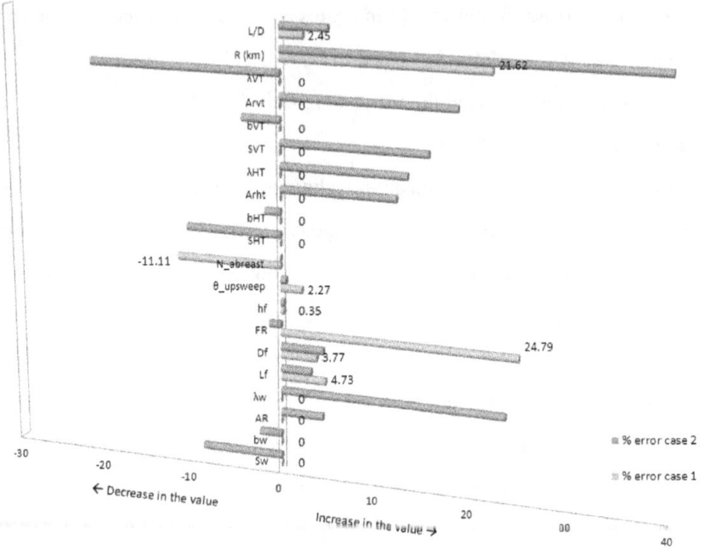

(a) Relative change in optimized design parameters

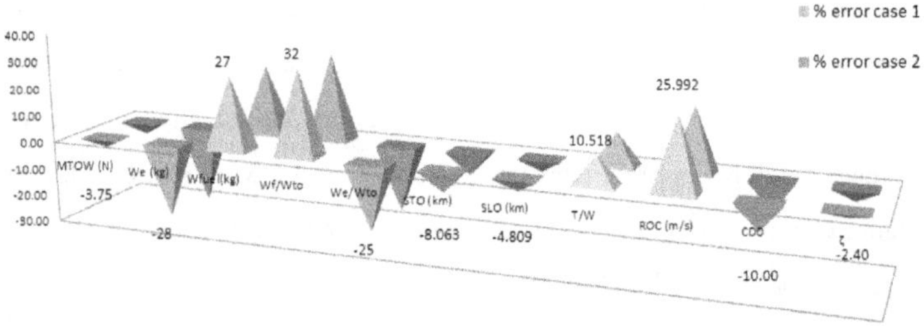

(b) Relative improvement in the objective function and its performance characterization between the cases

Figure 5.2: Graphical representation of variation in design variable and performance parameters variables is left open-ended and optimized by the drag minimum criteria. Fig.5.2b shows the comparison of different cases of performance, aerodynamics, and structural integrity of an airplane. It is evident that an overall MTOW is a decrease for case 2, and slight changes for case 1 are 4.841% and 3.75%, respectively. In scenario 2, the empty weight and fuel weight are lowered and raised by 26.268% and 17.144%, respectively, when looking at the fractional portion of MTOW. The results report that an increase in fuel weight may result in better performance and longer flying distances. Similarly, observing the empty weight fraction and fuel weight fraction factor, compared to baseline, decreases and increases in both cases, respectively. By considering performance in terms of take-off and landing distance, case 2 results superior to baseline and case 1. In Case 2, the take-off and landing distances have shrunk by 13.127% and 9.185%, respectively. Figure 5.2b highlights that the increase in the rate of climb leads to improvements in aircraft performance. For both cases, case 1 and case 2, the ROC has increased by 27.235% and 24.499%, respectively. In contrast to all the facts, for case 1 modifying the fuselage alone leads to improving the 0.233% of overall drag and hence improving other performance and structural considerations.To understand the dimensional variation of all the cases, a schematic overlap of three view diagrams is created using CAD software. Finally, an overlapping of the three-view diagram is presented for all cases manifested in Fig 5.3. A dimensional change in all the cases is properly highlighted in the front view, top view, and side view is represented in Fig 5.3.

Figure 5.2b illustrates the dimensional change in all the cases over the baseline model. Each of the cases is represented with a color code red, blue, and green to signify baseline, case 1, and case 2 respectively. The highlight of Figure 5.2a states that the decrease in wing span and increase in the vertical tail span is strongly notable as illustrated. This research extends the work to compute the error metrics among the methods. There are ample ways to predict the error to identify the degree of variance using percentage errors ([ ]). Concerning the level of error accuracy, Symmetric Mean Absolute Percentage Er-

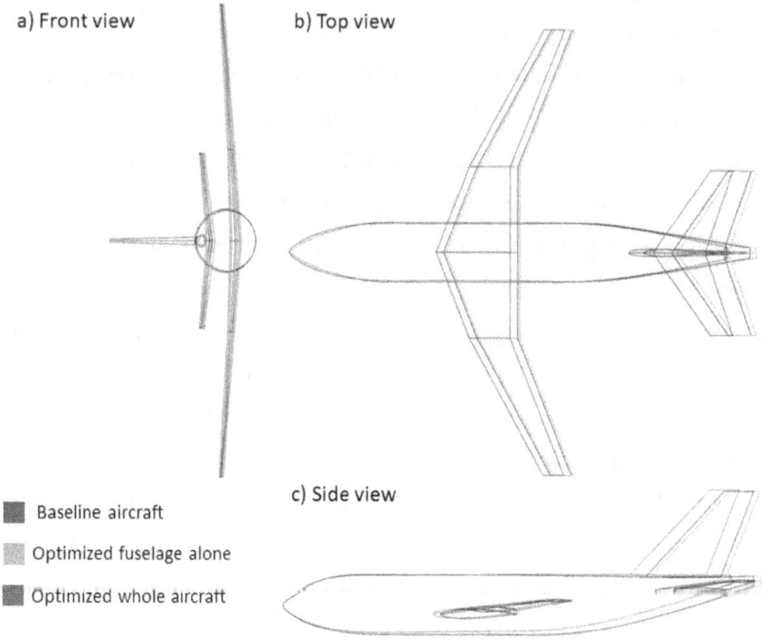

Figure 5.3: Comparison of all the cases with baseline model

ror (sMAPE) is used and expressed as in equation (13). Utilizing this equation to analyze the sensitivities of response to input uncertainty within a specific region can yield results that differ from sensitivity analysis conducted in the original design space. This kind of analysis is particularly valuable for aircraft vehicle designers. It not only highlights the relative significance of uncertain parameters but also guides efforts required to attain a desired level of success compared to the baseline vehicle. Where Aa is the values related to case 1 parameters, Oa is the other cases (baseline and case 2), and n is the number of design variables considered. (n = 1, 2,. . . ,13). Lewis [100] reported that sMAPE determines the accuracy of percentile value [[illegible]]. It states that if the percentile error is less than 10 percent seems to be strongly error-free. A slightly higher percentile of 10 to 20 percent indicates good accuracy. If the percentile value is within 20 to 50 percent valid for moderate accuracy and beyond 50 percent shows weak accuracy. In case a, the demonstration between case 1 and baseline indicates the sMAPE value 0.008 (0.833%), however, the error metrics disclose less than 10 percent which strongly defines error-free accuracy. In case b, a comparative study between case 1 and case 2 methods shows the error using sMAPe about 2.316%. Therefore the error metrics between case 1 and case 2 declared less than 10 percent show strong error-free accuracy. In case of c, a contrast between case 2 and baseline methods reveals that this combination shows an error metric of less than 10 percent strongly defines error-free accuracy. Thus, the outcome of the analysis shows the number of variables considered in different cases on the particular objective function is error-free. As a result, the error metrics show that case 1 is stronger at less than 10 percent proving to be strongly error-free accuracy in terms of considered design variable. Based on the results obtained, a detailed representation of the full-scale model was meticulously created using the NASA OpenVSP 3.24.0 version. This depiction aims to visually convey the precise arrangement and spatial distribution of its components, with the intricate details of the sketch presented in Figure 5.4. Moving forward into the subsequent chapter, there is an in-depth exploration of sensitivity analysis, specifically focusing on Case 1 and Case 2. This analytical inquiry seeks to understand how the model responds to variations in parameters or conditions, revealing the nuanced interplay between design factors and performance outcomes. The ensuing discussion will examine the implications of sensitivity analysis, providing valuable insights into the model's robustness and adaptability across different scenarios and configurations.

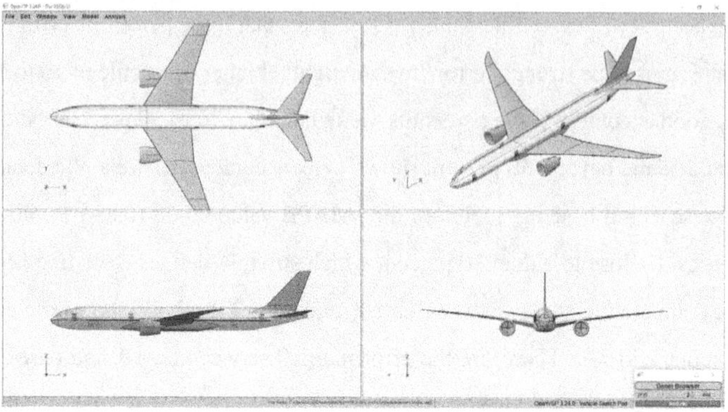

Figure 5.4: Overview sketch of proposed mode (case1) using OpenVSP 3.24.0

5.4 Future Implementation

The following module provides an overview of the managerial aspects of this article.

- An effective optimization strategy has created aerodynamic benefits for researchers, manufacturers, and executives, and delivers a highly precise and straightforward outcome with desirable input parameters [ ] [ ].
- This paper provides an advanced propulsive system with a high bypass ratio and low emissivity that can be utilized on the current operating aircraft to reduce emissions and improve fuel burn. Such concepts might lead to improving the range of a flight at a respective fuel.
- These optimized results clearly stated the human factors and passenger comforts inside the cabin, such outcomes and aspects are beneficial for any aviation manager, buyers, and the industry.

Chapter 6

Validation of a Novel Fuselage Design

6.1 Background

This chapter underscores the complementary roles of CFD, wind tunnel tests, and flight verifications in aircraft design. Wind tunnel tests are important, but adjustments and corrections are needed for scale effects and interference [ ]. CFD, while capable of full-size simulations without interference, still relies on wind tunnel data for validation, especially when flight measurements are not readily available during the design phase. This combination of tools ensures a thorough and accurate assessment of an aircraft's aerodynamic performance. In practice, wind tunnel data plays a crucial role in the design and testing process of modern aircraft. Wind tunnel data serves as a valuable benchmark for verifying the accuracy of CFD simulations, especially when flight measurements are limited [85][140][182]. This iterative approach ensures that the aircraft's aerodynamic performance is well understood and optimized throughout the design process. Furthermore, this chapter undergoes numerous studies that have shown that methodology adopted on various design factors significantly improves airplane performance. Numerous researchers have contributed the facts regarding validation approaches for new aircraft designs. In the broader context, some of the recent study is thoughtfully established and a comprehensive understanding of the research landscape serves as a vital foundation, ensuring a robust basis for implementing the innovative aircraft design tabulated in Table 6.1.

Many recent authors have explained various validation procedures, which are discussed in the sections that follow. A box wing configuration used iterative optimization to create a box wing configuration [ ]. Three different steps have been followed to go through this which include optimization of an airfoil, winglet height, and aspect ratio. The first phase resulted in a greater than 20% increase in lift-to-drag ratio over the baseline BAC-J box wing at 250 m/s cruise speed and a 5% increase through optimization of winglet height. Whereas the improvement through Aspect ratio optimization has resulted in little, with only a -5% aspect ratio model performed on par with the baseline [ ] [ ]. Thu et al. investigated a conceptual design code for a single-seat aerobatic Airplane (SSA) that was built and validated which falls under the Multidisciplinary design optimization technique (Numerical optimization). For solving nonlinearly confined optimization problems, Sequential Quadratic Programming (SQP) has been employed. The SSA conceptual design code was effectively developed and evaluated against existing SSA. For better performance, the verified SSA has been used as a baseline model to optimize for the lowest possible overall Drag coefficient. The optimization reduces total drag while simultaneously reducing take-off distance, increasing maximum range, and increasing the Rate of Climb [ ]. Dakka et al., [ ] studied about wings and fuselage are combined to create a single lifting body which can be called a BWB (Blended Wing Body) aircraft. For this, an aerofoil NACA 0012 has been considered because it is ideal for subsonic conditions. To decrease manufacturing costs, the design was tested in Ansys Fluent (CFD) to learn about the behavior of BWB [ ]. The results proved that there is an increase in L/D than conventional aircraft. Additionally, such concepts provide better aerodynamic performance at high subsonic speeds than at low subsonic speeds. In this work, the proposed work concludes that BWB has performed 9.4% better than compared to a conventional aircraft

Table 6.1: Literature survey on Aircraft design validation approaches

Applied	Main findings	Methodology	References
Transport Aircraft	Efficient correlation study between CFD-wind tunnel methodology	CAE–AV model and its test in the German-Dutch Wind tunnels (DNW).	[85]
Transport Aircraft	To create an efficient and more profitable single-seater airplane	Optimized using SQP optimizer of non-linear program (NLP).	[173]
Box Wing Concept Aircraft	To improve the aerodynamic efficiency by optimizing the L/D, AR and winglet height.	Implemented CFD, and Solid Works.	[98]
BWB Aircraft	To improve aerodynamic Characteristics at low to medium subsonic speeds.	Conducted CFD simulation and validated through Wind tunnel testing techniques.	[97]
Fighter Aircraft	To increase in lift force by up to 12% by implementing canard wing and leading-edge extension.	Conducted CFD simulation and flow visualization and aerodynamic forces at Wind tunnel.	[188]
Sustainable aircraft design	to improve the power Loading by multi-optimization techniques.	Implemented a MDO with NLP Conducted CFD simulation	[137]
New Aircraft	a shape optimization of a BLI intake	Utilizing a MOGA and validated with CFD	[168]
Conceptual Aircraft Design	Minimized the weight by 3% of a novel aircraft by MDO approach	Applied MDO approach and computed CFD analysis	[114]
This Research	Modified rear fuselage as design optimization as Objective function to Enhance performance	Optimized with non-linear program (NLP), and validate computational and Experimental results.	-

[ ]. Wibowo et al., [ ] study of a canard in front of the wing is also a technique to increase the efficiency of an aircraft by guiding the flow along the fuselage. It also acts as a stabilizer, and an elevator and increases the lift of an aircraft. In the current study, it has been proved in Analysis tests that adding canard can increase the lift coefficient up to 12% [ ]. Further extending, Peter et al., [ ] provided an overview of the key modeling approaches and optimization strategies for designing new upcoming hybrid-electric aircraft. In addition, an easy-to-understand optimization model was offered as a reference showcase to exemplify the potential provided by applying optimization and sensitivity analysis approaches in the conceptual design phase of hybrid or fully electric aircraft. Sudasinghe et al., [ ] introduced the boundary layer ingesting (BLI) propulsion system, which refers to the ingestion of the aircraft wake by the propulsors, has received a lot of attention and has been linked to improved fuel economy. To decrease the pressure losses and distortion in the engine inlet, the shape of BLI intake has been optimized by using combined techniques, free-form deformation, and a multi-objective genetic algorithm. Cui[ ] worked to evaluate the aerodynamic performance of the latest BWB design with 450 seater and assess potential engine integration layouts. Comparing Computational Fluid Dynamics (CFD) with experimental data yielded favorable agreement. Liu et al., [ ] investigate a traditional aircraft design procedure, the horizontal and vertical tails are typically sized using volume coefficient techniques. This publication presents an enhanced method for conceptual aircraft tail design which is based on a multidisciplinary design optimization (MDO) strategy with stability and control constraints. A test case for a concurrent wing and tailplane design is presented, and the results show great potential for fuel burn and carbon reductions with empennage design optimization at the conceptual aircraft design phase. The test case result has reduced more than 9% in aircraft block fuel weight and reduced by more than 3% in aircraft maximal take-off weight [ ]. Upon a comprehensive review of recent studies focused on validating aircraft designs, it becomes evident that a robust validation strategy necessitates the integration of

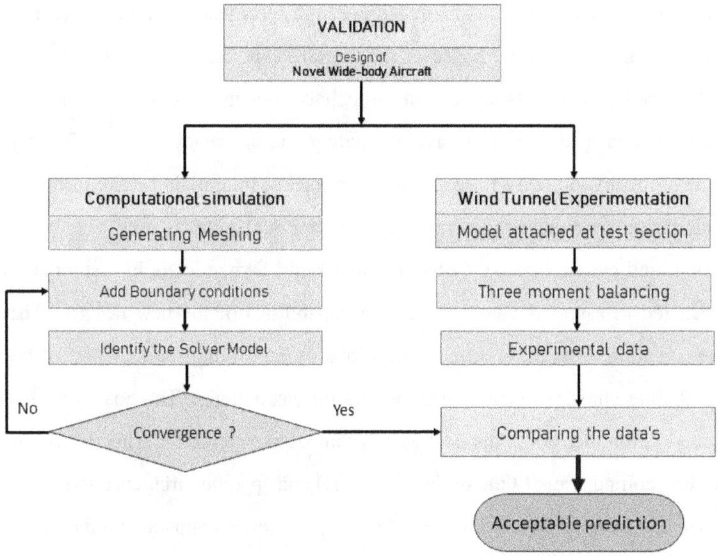

Figure 6.1: Validation approach for the novel aircraft design

Computational Fluid Dynamics (CFD) techniques, contingent on the specific flow regime, in conjunction with wind tunnel testing to corroborate and validate the obtained results [ ]. Subsequently, the forthcoming section in this study delves into an elaborate exposition of both CFD approaches and the wind tunnel methodology, and concluding remarks from the validation process as illustrated in Figure 6.1.

6.2 Computational Approach

Traditional aircraft design methods are applied to develop and evaluate the aircraft, with a particular emphasis on design requirements and studies covering weight sizing, pilot ergonomics, aerodynamics, stability, and performance. The primary focus is on optimizing various aerodynamic factors to achieve high aerodynamic efficiency (L/D), which is critical during the working phase of the aircraft. The research emphasizes the influence of aircraft components and advanced technologies in achieving these goals. Wake turbulence is a critical concern in aviation safety and efficiency. This CFD analysis focuses on a novel wide-body fuselage design aimed at minimizing wake turbulence in a subsonic passenger transport aircraft. The Airbus A310-200 is used as a baseline for validation, and comprehensive Computational Fluid Dynamics (CFD) simulations will be conducted to assess the wake-reduction capabilities of the new design. The study involves creating three different configurations by modifying the rear part of the fuselage, including changes to cabin spacing and upsweep angle. The goal is to determine the configuration that offers the best aerodynamic characteristics. This determination is made using computational fluid dynamics (CFD) and specific turbulence models, such as the Reynolds–Averaged–Navier–Stokes (RANS) equations coupled with the Shear Stress Transport (SST) model. The evaluation includes assessing lift, drag, and pitching moment coefficients at various angles of attack, and of the different configurations. The research aims to demonstrate how these configurations hold promise for enhancing aerodynamic efficiency and improving the performance of the novel aircraft aircraft, as highlighted in the CFD study. The CFD process recommends these key investigations as a standard baseline for plausibility analysis:

- Visualize y+ distribution around all bodies, ensuring appropriate cell values.
- Examine pressure distribution around the fuselage, validating its conformity to expectations.
- Assess velocity near the fuselage, ensuring disturbances decrease smoothly away from it; sudden changes may indicate inappropriate discretization.
- Check stagnation conditions, pressure, and velocity locations, with a non-zero stag-

nation velocity signaling poor discretization.

- Monitor the wake region propagation, aiming for smooth integration; abrupt changes may suggest inappropriate discretization.

6.2.1 CFD Simulation Steps

In-depth CFD simulations are conducted to assess the aerodynamic performance of the novel design. These simulations evaluate lift-to-drag ratios, drag coefficients, and airflow patterns to validate its improved efficiency compared to conventional designs as follows:

- Geometry and Mesh Generation: Detailed 3D models of the novel design and the Airbus A310-200 are created, and high-quality computational meshes are generated to capture intricate details and flow characteristics. The construction of an unstructured polyhedral mesh, coupled with refinement regions, exhibited the closest agreement with experimental values. As a result, all subsequent simulations will adopt an unstructured polyhedral mesh along with comparable mesh refinement regions [131] adopted using Ansys 19.1. An in-depth study of the meshing of a scaled model is performed to check the robustness of the computational analysis by mesh independence study [106]. The study of mesh independence provides a clear justification for the outcome obtained from the mesh created and the outcome achieved. In this research, initially model is imported in ICEM, then using an online y+ calculator set the boundary constraints are illustrated in Table 6.2. Based on the previous research mesh analysis, variations of mesh density are created and re-analyzing the output until to achieve a convergence rate. The variation of mesh density shows that after the number of mesh elements is 6050097, the drag force value remains approximately the same as shown in figure 6.2. The final meshing of

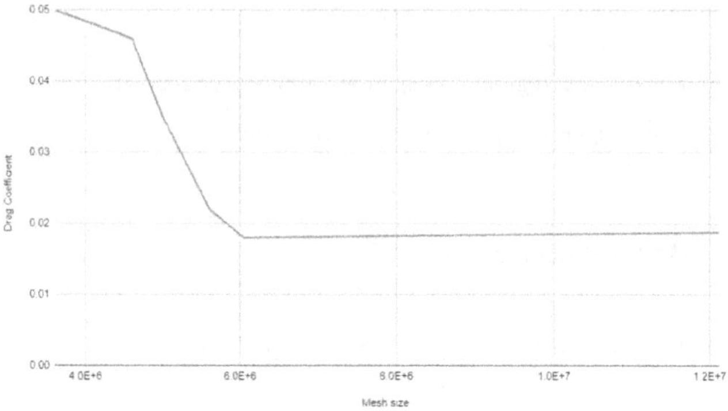

Figure 6.2: Grid independency

Table 6.2: Meshing Constraints

Global Mesh	Element Size:100
Mesh type	Tetrahedrons
Mesh Algorithm	Patch conforming

the model is illustrated in the figure 6.3. Further, using a final meshing module the mesh is generated as shown in the figure 6.5.

- Inlet Conditions: Realistic flight conditions, including airspeed, altitude, and environmental factors, are set up for accurate simulations.

- Turbulence Modeling: On several research papers, both Reynolds-Averaged Navier-Stokes (RANS) and Large Eddy Simulations (LES) are conducted to analyze wake turbulence under varying flight scenarios. Appropriate turbulence models are employed to capture wake turbulence dynamics accurately. Within the realm of Computational Fluid Dynamics (CFD) software analysis, an intricate examination of fluid flow and its pertinent physical attributes—namely, velocity, pressure, viscosity, density, and temperature—is conducted. This computational analysis hinges on the meticulous definition of operating conditions. Notably, the CFD testing em-

Figure 6.3: Meshing of Model

ploys the k-omega SST model, a turbulence model widely acknowledged and accepted as an industry standard [[illegible]] [[illegible]]. Derived from insights gathered during a comprehensive literature review, the k-omega SST model emerges as a robust choice, showcasing superior predictive capabilities for flow separation when compared to most Reynolds-Averaged Navier-Stokes (RANS) models. The outcome of the pressure distribution and velocity distribution over the surface is shown in figure 6.4.

The primary goal of the novel wide-body fuselage design is to minimize the drag of the aircraft. The fuselage design integrates aerodynamic features, smooth contours, and tailored flow control devices to reduce the adverse drag effects Post-processing tools visualize and compare wake structures, vortex behavior, and wake decay between the novel design and the Airbus A310-200. Wind tunnel tests validate the proposed model's aerodynamic properties and stability. Data collected from these tests confirm its drag reduction capabilities and highlight any design refinements required for optimal performance. Ef-

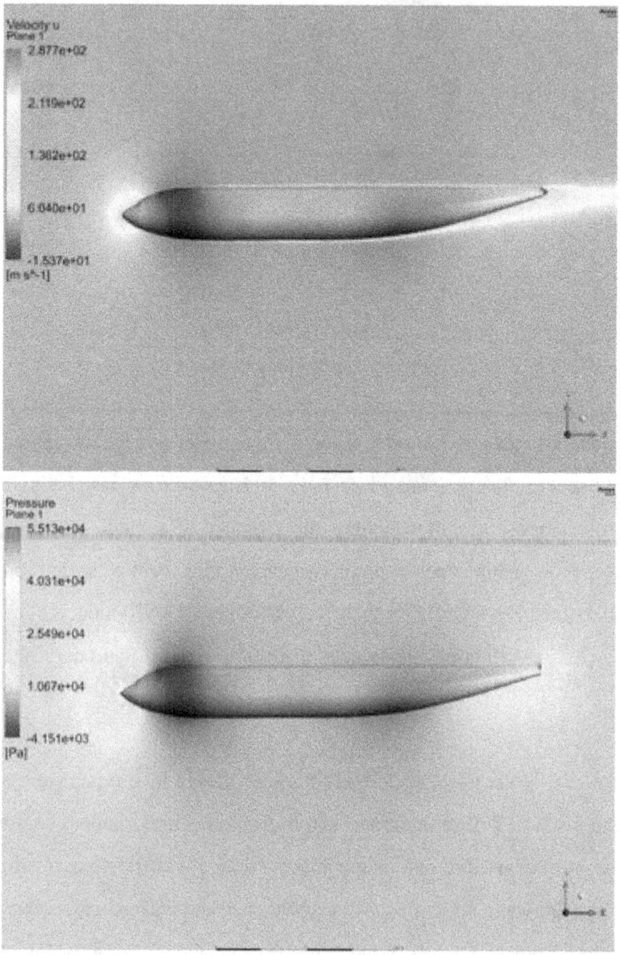

Figure 6.4: Analysis of Pressure and velocity distribution over a fuselage surface

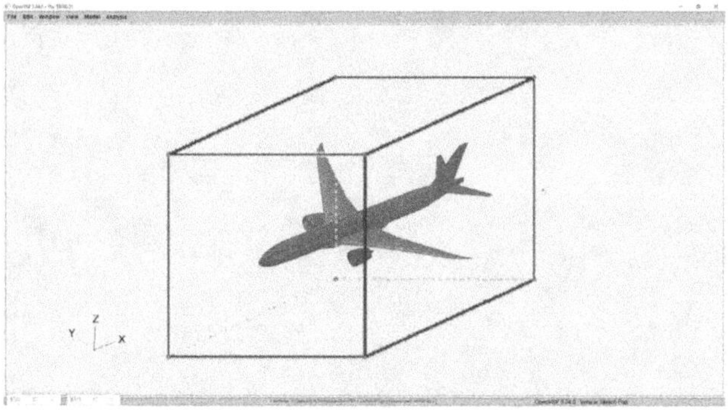

Figure 6.5: Meshing of the proposed model

forts to enhance aircraft efficiency and reduce drag are essential for aviation progress. This study combines wind tunnel experiments and computational analysis to investigate a novel wide-body fuselage design aimed at minimizing drag from the rear part of the aircraft. By using the Airbus A310-200 as a reference point, the validation process compares the proposed design's drag characteristics with real-world and simulated data.

6.2.2 Boundary conditions

As the modification of baseline aircraft mainly focuses on the fuselage the 2D analysis is carried out to understand the different types of flow parameters and pressure variations on both fuselages and are compared. The process is started by importing the 2D model into the Ansys design modular. The domain is created in a modular design using sketching tools. Further, using a meshing module the mesh is generated as shown in the figure 6.5.

The same process is followed for both the fuselage and rear parts. The meshed model is imported into the fluent solver with the given inputs mentioned in Table 4.4. After the initialization of the solver is done. the calculation of the model is initiated with a 1000-iteration process. The solution is converged for the A310-200 fuselage rear at the 455th

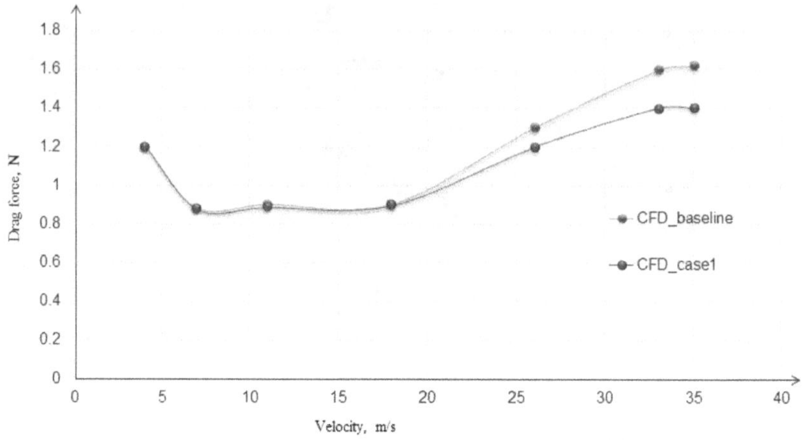

Figure 6.6: CFD variation between baseline and proposed model

iteration and the modified aircraft 652nd iteration. The solutions were obtained in the form of graphs and contours. The pressure contours of both fuselage rear parts are shown in Figures 6.7 and the pressure variation graph is obtained from the Ansys fluent solver.

6.3 Wind Tunnel facility

This chapter utilized wind tunnels for testing prototypes where throughout a test run, air is passed over the prototype. Where the physical models of the Airbus A310-200 and the proposed design are created with precision to replicate geometry and details. Detailed 3D physical models of the novel design and the Airbus A310-200 are constructed, ensuring accurate representation of geometries and details. The physical models are placed within a wind tunnel equipped to simulate flight conditions. Inlet conditions including airspeed, which relates, and turbulence levels are set according to real-world scenarios [ ] [ ]

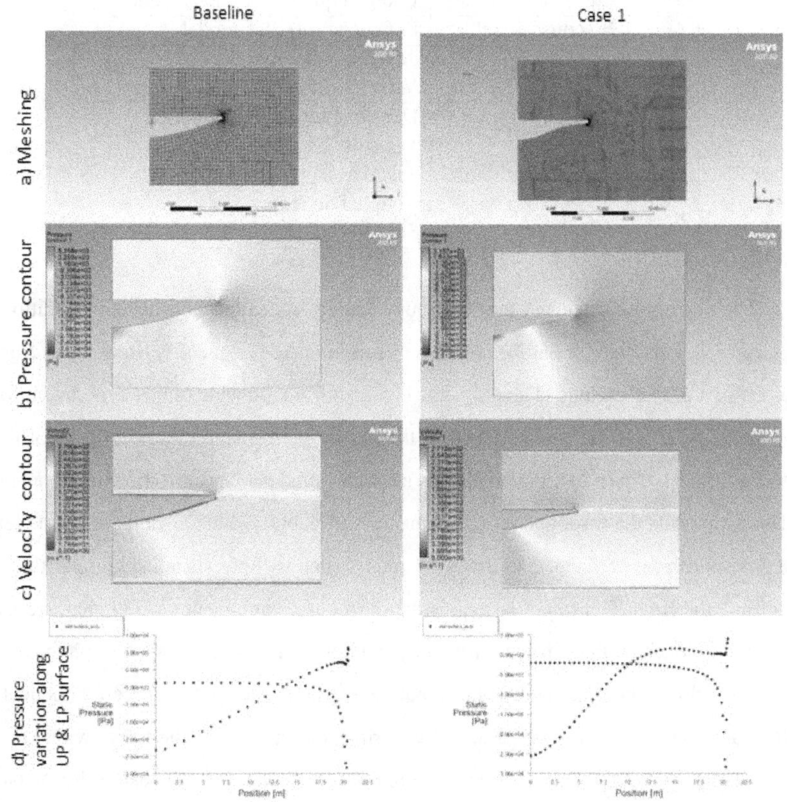

Figure 6.7: Comparative CFD Analysis

Table 6.3: Boundary condition of the input value for the solver

S no.	Input parameters	Input values
1.	Solver method	Implicit method
2.	Solver Type	Pressure Based
3.	Equations selected	Energy equation, Viscous model
4.	Viscous Model	k- Omega (ω) SST model
5.	Inlet conditions	Taken as velocity inlet Velocity: 253 m/s
6.	Inlet Properties	Density : Ideal Gas Viscosity Model: SutherLand
7.	Outlet conditions	Taken as a pressure outlet
8.	Methods	Coupled Scheme Second order upwind

[ ]. The models are subjected to wind tunnel testing, with adjustable airflow conditions emulating various flight scenarios. Measurements are taken to quantify drag values from the rear part of the fuselage. The drag data obtained from the wind tunnel experiment for the Airbus A310-200 is validated against the known performance characteristics of the actual aircraft. Intricate instrumentation as a balance module measures the three-moment balance capturing the aerodynamic performance and stability value to validate the model. Many tests can be performed in the wind tunnel such as flow visualization and three-moment balancing. The force balance is the most basic equipment used to characterize the six components: three forces (lift, drag, and side) and three moments (pitch, roll, and yaw). Some three-moment balance equipment is used to assess three components (lift, drag, and pitch). The wind tunnel experimental analysis of the novel wide-body fuselage design seeks to validate its wake-reduction capabilities. By using the Airbus A310-200 as a benchmark, the study aims to provide empirical evidence of how the new design minimizes wake turbulence. Successful validation could lead to innovative design strategies, contributing to enhanced aviation safety and efficiency in wide-body subsonic passenger transport aircraft. A comparison of aerodynamic characteristics between the novel design and the Airbus A310-200 serves as the validation process.

6.3.1 Wind Tunnel Specification

The main objective of this chapter is to determine the aerodynamic forces of the baseline model and novel aircraft model. Both the aircraft model is made of 3D printing technology of 0.256 mm accuracy. The novel aircraft model is tested in open subsonic wind tunnel tests at IISC Bangalore, with a 120 cm long, 60 cm X 60 cm cross-section area of test-section as shown in Figure 6.8. In this research, consideration of this wind tunnel is preferred as it is dedicated to the simulation of aircraft models, and airfoil structures under subsonic limited to 70m/s and turbulence intensity about 0.10%. During the wind tunnel experiment, the flow speed is adjustable from 5 to 50 m/s.The scale ratio of the model was 1/100 of the full scale, and the similitude Reynold's model is followed [108] [illegible]. The work has been followed as per the precautionary act before running the Wind Tunnel:

- Ensure no one stands behind the motor during operation.
- Before starting, check the fan area and only allow competent personnel to handle the controller. Secure main power.
- Prior to tunnel startup, inspect the test section for loose parts, removing any found. Secure the test section with C clamps.
- Avoid running the tunnel below 100 r.p.m. to prevent motor overheating. Intermittent running at lower speeds is acceptable, but not for extended periods.
- Limit prolonged operation at higher speeds whenever possible.
- Ensure the model is well connected with the module settings. Ensure no nuts or other components are loosely connected during these checks.

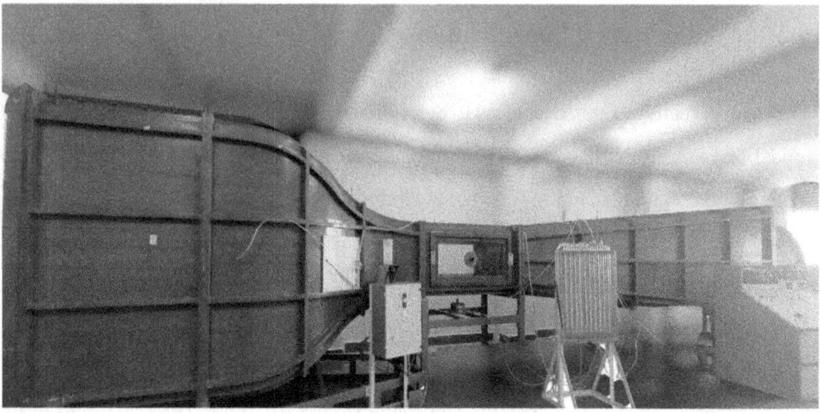

Figure 6.8: A wind tunnel

6.3.2 Wind Tunnel Model

Fabrication of the baseline model aircraft and modified model aircraft has been done using 3D printing. 3D printing is also widely known as additive manufacturing which is a technique of manufacturing by adding material. This process starts with the creation of a computer-aided design model using different types of CAD software. The baseline aircraft and modified aircraft have the same rear fuselage section percentage (44% of the whole fuselage length). There are various types of technologies and processes are available nowadays. According to the requirements and specifications type of process and technology is chosen. In this research work, Fused deposition modeling (FDM) is a liquid-based additive manufacturing technique used for fabrication. The material used in this technique is Polylactic acid (PLA). PLA is chosen because it is eco-friendly, cost-efficient, and has a high strength-to-weight ratio of 3.29 whereas carbon fiber has a strength-to-weight ratio of 3.58 as shown in Figure 6.9. Thus, compared with PLA carbon fiber is costlier. Furthermore, the model scaling is one of the key parts while testing

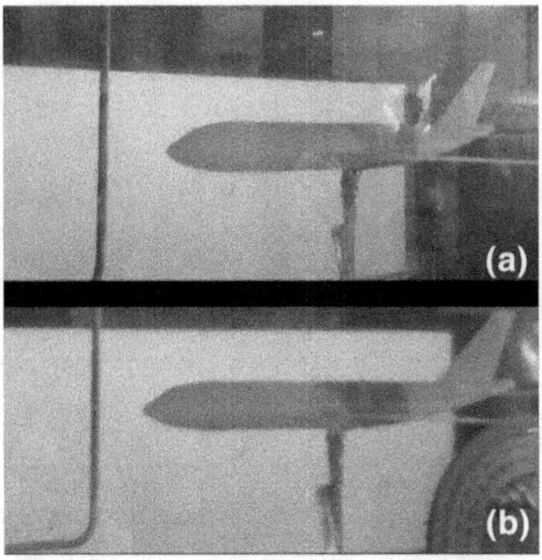

Figure 6.9: Scaled model (a) baseline (b) optimized (case 1)

in the wind tunnel. Understanding the wind tunnel constraint test section sizing, and the concept of solid wall blockage, the base and novel models are scaled down to 1/100th, accordingly.

6.3.3 Measuring techniques

In this research, the three-moment balance equipment is used to obtain the lift, drag, and pitching moment values of the model placed in the wind tunnel [ ] [ ]. In this experiment, the focus was on a force balance system that enables the direct measurement of critical aerodynamic parameters, specifically lift, drag, and pitching moment as shown in Figure 6.10. This equipment relies on load cells or force transducers to achieve these measurements effectively. Such experiments extend their capabilities to encompass the measurement of the aerodynamic forces and moments, providing a more comprehensive understanding of the aerodynamic behavior of the tested object in the wind tunnel. Utilizing the three-moment balance methodology, the primary objective is to anticipate the drag

Figure 6.10: Three Moment balancing equipment

forces acting on both the conventional baseline aircraft model and the scaled-down novel proposed model. These experimental models undergo rigorous testing within an identical wind tunnel, maintaining consistent Reynolds numbers and wind tunnel configurations throughout the experimental procedure. The obtained results are then presented graphically, illustrating the variation in drag forces concerning the increase in velocity and its corresponding impact on the aircraft models as shown in Figure 6.11. The graphical representation unveils that the novel proposed aircraft model exhibits a notable comparative advantage, demonstrating superior performance for two fundamental reasons. Firstly, the proposed model experiences a minimal delay in drag compared to the baseline model. Secondly, at an rpm (revolutions per minute) value of 999, the drag force encountered by the proposed model is significantly lower than that experienced by the baseline model.

In summation, the findings conclusively indicate that the proposed model boasts a more favorable aerodynamic profile in terms of drag force when contrasted with the baseline model. This is primarily attributed to the reduced drag delay and lower drag values observed at specific rpm settings in the proposed model. Consequently, this chapter substantiates the validity of the research objective, affirming the superior performance of the proposed aircraft model in mitigating drag forces.

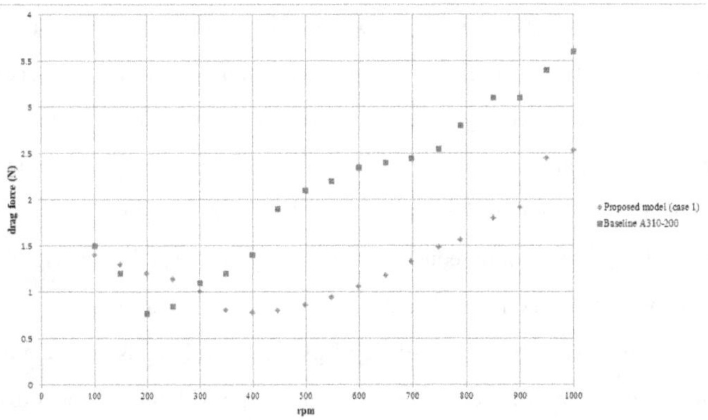

Figure 6.11: Comparative wind tunnel testing between baseline and proposed model

6.4 Validation and Comparison

The main aim of this research work is to improve performance by modifying the fuselage which results in improved aerodynamic efficiency. To observe these results CFD analysis has been done on the fuselage using Ansys R 2022. Following the viscous model, large eddy simulation (LES) is suggested in the research papers as the optimum turbulent model to analyze a fuselage. An inlet is taken as a velocity inlet with a velocity of 30 m/s. From the obtained CFD data over the rear fuselage of both A-310 200 & modified aircraft, it can be concluded that the stability of the aircraft is increased as the increased positive pressure values for the bottom part modified fuselage rear end compared to the A-310. this gives the possibility of fixing the engine at the rear end of the fuselage which is one motive of this project. The velocity at the rare part of the modified fuselage is more compared to the A310-200 which allows us to fulfill another objective of this project which is to increase the intake of the aft-mounted engine by boundary ingestion concept. Figure 6.7 explains the pressure variation along the upper and lower ends of both fuselages' rear ends. The drag values obtained from wind tunnel experiments for the Airbus A310-200 are compared with the results of computational analysis. Any disparities are analyzed to ensure the accuracy of both approaches. Comparison of Proposed Design: Drag force values obtained from computational analysis of the proposed design are compared against the validated results of the Airbus A310-200 as shown in Figure 6.12. This demonstrates the potential drag reduction achieved by the novel design. Through the validation of wind tunnel experimental analysis with computational analysis, this study aims to confirm the drag-reduction capabilities of the novel wide-body fuselage design. By comparing the drag performance of the proposed design with the validated results of the Airbus A310-200, the study contributes to a better understanding of the design's effectiveness in minimizing drag from the rear part of the fuselage. According to the American Statistical Association, the concept of error analysis quantifies uncertainty by applying mean squared error. In this case, analysis between the wind tunnel test data and cfd values, 2.9379 is obtained which is within the reasonable range. The value of mean squared error shows a successful validation of the novel design by the adoption of innovative design strategies for improved efficiency in subsonic passenger transport aircraft. Figure 6.7, explains the pressure variation and velocity contour along the upper and lower ends of both fuselages' rear ends. The computation result shows improvement in high pressure at the lower rear end of the fuselage which will help to create nose down moment. In the point of stability and control of the flight, this shows a positive impact on the stability and balancing of the systems. Whereas, the velocity contour shows the deceleration of velocity that proves a reduction of wake at the rear part of the proposed design.

Key Findings

• The proposed design involves a high camber at the rear section leads to increased high pressure at the bottom rear end of the fuselage and incorporating boundary layer ingestion

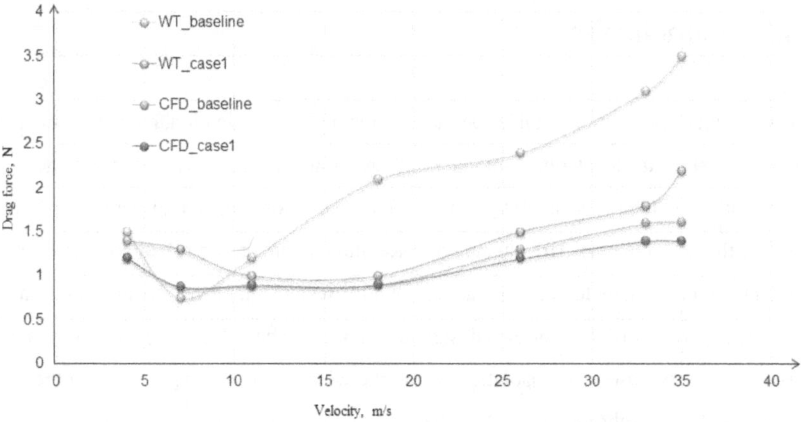

Figure 6.12: Comparative result between wind tunnel testing and CFD outputs

helps to reduce leakage at the rear part of the fuselage.

• The proposed section increases high pressure at the bottom rear end of the fuselage and results in lower velocity in it.

• Looking at the balancing and stability point of view, which provides pitch down moment. Adds benefits from the stability point of view of the proposed design.

• Flow visualization provides prominent results and reveals the effectiveness of the design to reduce wake at the rear end of the fuselage.

Managerial Insights

• The research can extensively improve the market size of production and profitability.

• Many global programs took the initiative to improve aerodynamic efficiency which is the direct and indirect pathfinder towards sustainability.

6.5 Conclusion

This chapter systematically undertook a comprehensive examination, encompassing both wind tunnel testing and Computational Fluid Dynamics (CFD) analysis, for both the baseline model and the proposed design model. The evaluation aimed at discerning and comparing the aerodynamic performance of these models under controlled conditions. The wind tunnel testing yielded empirical data, and the results unequivocally indicated a noteworthy superiority of the proposed design model over the baseline model. This superiority was not merely observed in isolation within the wind tunnel setting but was further corroborated by the subsequent CFD analysis. The consistency of the findings across both experimental methodologies reinforces the credibility and robustness of the observed improvements in the proposed model. Therefore, based on the congruent evidence from both wind tunnel and CFD analyses, this chapter draws a firm and conclusive inference: the proposed design model outperformed the baseline model in terms of aerodynamic characteristics. The convergence of positive results from two distinct methodologies adds weight to the assertion that the proposed model indeed represents a superior and more effective design when compared to the baseline model.

According to the results obtained, it can be concluded that the expected results are well validated through CFD and Wind Tunnel results. Even though, this novel work can be further modified of the fuselage with the implementation of riblets and a combination of both the boundary layer ingestion (BLI) and riblets for more increment in the aerodynamic efficiency. Similarly, the input conditions are based on the defined objective. Such novel work can be utilized in Smoke flow visualizations in order to further evaluate the fuselage profile using a low subsonic wind tunnel.

The article expresses the work's potential for expansion in two ways: To begin, parametric analysis may be used to compare the robustness of the model. In addition, comparing the mix of technologies or incorporating enhanced retrofit designs. Second, both computational and experimental work may be done to characterize the flight's performance characteristics. Following successful flight testing, innovative aircraft can be certified for improved airline profitability and fuel economy. Third, as the research study indicates, alternative interdisciplinary design optimization models may be conducted by other parameters. Furthermore, the challenges in obtaining data, some factors are examined, and the interactive parametric research is not considered. In future studies, the suggested model might be combined with other theories to improve performance effectiveness.

Chapter 7

Conclusions, Limitations, and Future Scope

7.1 Conclusion

Nowadays, the aviation industry stands at a crossroads and seeks to strike a balance between global connectivity and environmental responsibility, this approach represents a beacon of hope. This synergy finds its embodiment in the conceptual design of sustainable aircraft—a realm where cutting-edge technology and ecological consciousness converge to shape the future of flight. This research aims to modify the fuselage design based on the Middle-of-the-Market (MoM) aircraft and stands as a pivotal opportunity for the aviation sector to take another historic leap. As the aviation industry embraces the challenges and possibilities of the next decade, the conceptual design phase takes center stage, laying the groundwork for a new era of wide-body commercial aircraft that seamlessly combines innovation, efficiency, and passenger-centric design. This elaboration delves into seven diverse aircraft design concepts: Double Bubble Aircraft, Joint or Box Wing, Blended Wing Body (BWB), Flying V Aircraft, Transonic Truss-Braced Wing, Propulsive Fuselage concept by Bauhaus Luftfahrt, and Strut-Braced Wing. Each case study illuminates how these designs leverage distinct approaches to enhance fuel efficiency while addressing common challenges.

This research project commences with the comprehensive real-time collection of data from wide-body aircraft. A meticulous investigation is conducted, encompassing aspects such as aerodynamics, structural integration, operational performance, and passenger comfort. The primary focus of this study is to explore these facets in great detail. Subsequently, through a meticulous case study, a thorough examination of the interrelationships among these parameters is performed, guided by findings from existing literature. The selection of these parameters is based on their substantial correlation coefficients, which signify their pivotal roles in understanding various aspects of wide-body aircraft performance. Of all the parameters scrutinized, the Maximum Takeoff Weight (MTOW) stands out as a critical factor, being identified as the root cause of numerous challenges arising from increased MTOW. The literature review highlights various challenges posed by high MTOW and offers insights into potential strategies for weight reduction in aircraft design.

Moving forward, the research confronts the task of choosing a baseline aircraft model that aligns with market requirements. To address this complex decision-making process, a mathematical technique known as Multi-Criteria Decision Making (MCDM) is employed. Among the MCDM methodologies, the CRITIC-WASPAS approach is favored over the Technique for Order of Preference by Similarity to Ideal Solution (TOPSIS). This approach aids in identifying the most suitable baseline aircraft, ultimately leading to the selection of the Airbus A310-200. This aircraft serves as a foundation for further modifications that can address the needs of the aviation market more effectively. With the baseline aircraft established, the research transitions into the realm of design optimization. The primary objective function for this optimization process is the MTOW. Design variables are systematically adjusted within the constraints stipulated by airworthiness standards. A critical aspect of the research involves assessing the robustness of the objective function, particularly in comparison to a case denoted as Case 1. Case 1 excels in multiple dimensions, particularly in enhancing aerodynamic efficiency, rate of climb (RoC), take-off distance, and landing distance.

To validate the results obtained from the design optimization, a physical model is fabricated for wind tunnel testing, and computational simulations are carried out using Computational Fluid Dynamics (CFD) software. The analysis of the data from these two sources reveals a modest 11% margin of error, indicating a high level of compatibility and confidence in the research findings. In summary, the innovative fuselage design demonstrates its superiority over the baseline Airbus A310-200 model across various performance metrics. This conclusion underscores the potential of the modified design to offer enhanced performance and operational capabilities in the aviation market. As a closing remark, it's worth noting that this research has the potential for further expansion, allowing for the exploration and prediction of various flight characteristics and their impact on the aviation industry.

7.2 Limitations and Future Scopes

This research work exhibits several limitations, which, however, provide valuable opportunities for dedicated researchers to explore and improve further. This section will comprehensively outline and discuss these limitations, acknowledging their presence while highlighting the potential avenues for future investigation and enhancement:

- This research work presents a specific focus, concentrating primarily on the examination of fuselage design configurations and the influential factors impacting passenger comfort in relation to seating arrangements [117]. However, the scope of this study invites future researchers to delve deeper into the intricacies of aircraft design, expanding the analysis to encompass the holistic evaluation of wing-empennage-fuselage combinations.

- Within the context of this research, our investigation is limited to the passenger seating arrangement, with a view to future studies that could undertake a more in-depth exploration of novel aircraft passenger ergonomics. Such an investigation holds the potential to yield insights crucial to enhancing the overall aviation industry's profitability.

- This research work was performed based on the limited design variables and parameters. Furthermore, the study can be extended based on the various other design parametrical studies that can be performed and study the influential effect on the performance of aircraft. Expanding the research to encompass these additional variables will enable a more comprehensive understanding of their influence on aircraft performance. Moreover, a comparative study and sensitivity analysis of these parameters can shed light on their respective effects on aircraft performance, adding depth to our knowledge in this area.

- In this study, the Hybrid MCMD (Multi-Objective Constrained Multi-Disciplinary Design) method and interior point optimization technique were employed to minimize the Maximum Takeoff Weight (MTOW) within the constraints of specific design variables. Future research endeavors can extend the investigation by varying the objective function. This variation could encompass examining parameters such as drag coefficient, fuel burn, emission index, and other performance indices, offering a more comprehensive assessment of an aircraft's overall performance characteristics.

- It is worth noting that the current research primarily focuses on analyzing the performance characteristics of the aircraft. Yet, there exists a promising avenue for expansion, wherein the study can evolve to encompass the dynamic flight dynamics characteristics and potentially move into the realm of prototype flight testing, even within a controlled environment like a vertical wind tunnel. This extension could offer invaluable insights into the practical application and real-world performance of the designed aircraft, further enriching the body of knowledge within the field of aviation research.

Appendix

A1. Code for Objective Function

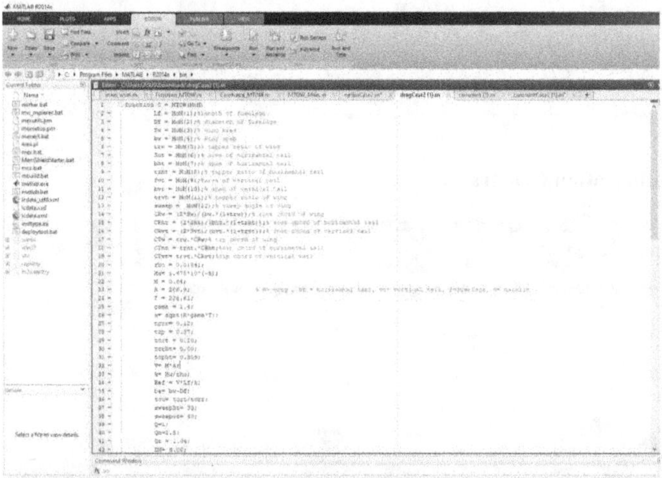

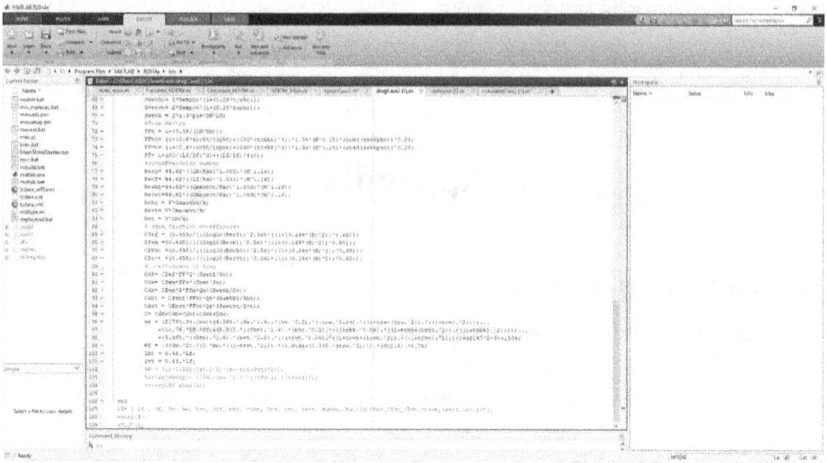

A2. Code for Optimization Constraint

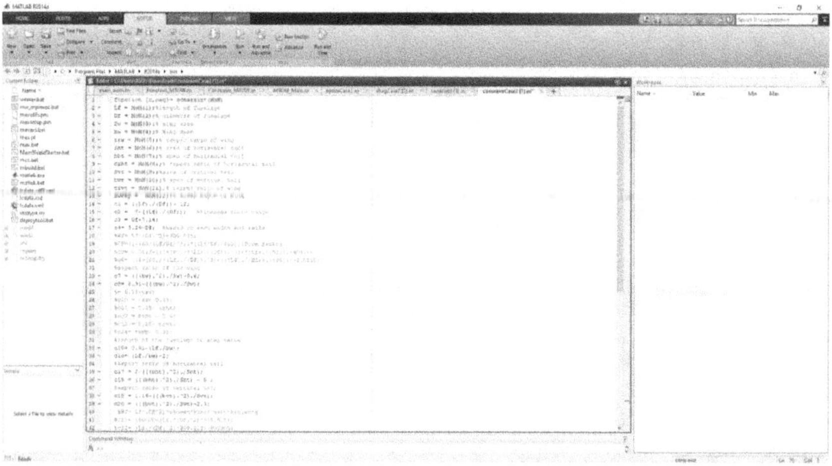

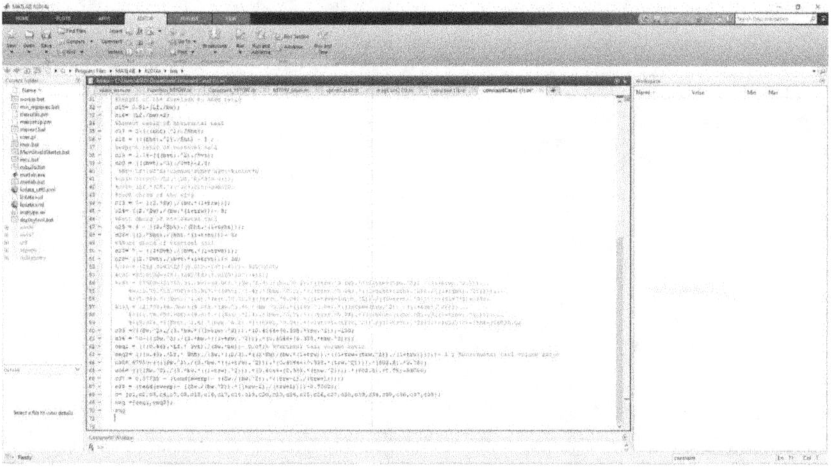

A3. Code for Optimization Solution

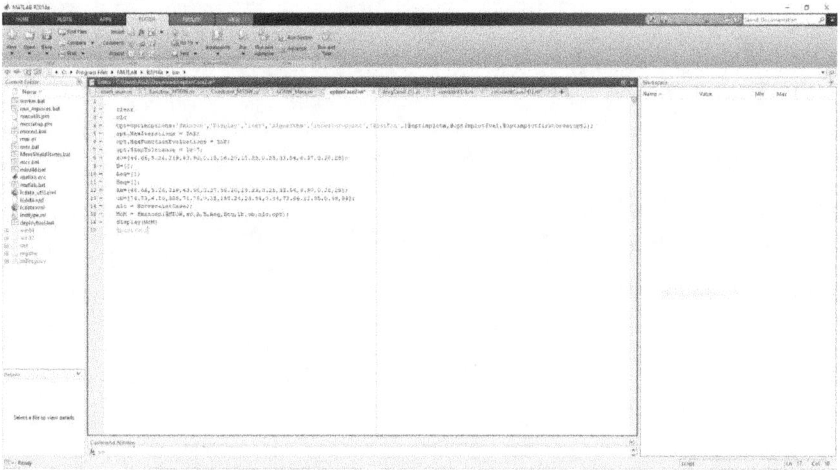

www.ingramcontent.com/pod-product-compliance
Lightning Source LLC
LaVergne TN
LVHW010727061126
851857LV00014B/1379